CURIOUS MINDS

John Brockman is founder and CEO of Brockman, Inc., an international literary and software agency; president of Edge Foundation, Inc.; publisher and editor of *Edge*, a Website presenting the third culture in action; and co-founder of rightscenter.com, Inc., an Internet company. He is the author or editor of four books about science.

ALSO BY JOHN BROCKMAN

As author

By the Late John Brockman
37
Afterwords
The Third Culture: Beyond the Scientific Revolution
Digerati

As editor

About Bateson
Speculations
Doing Science
Ways of Knowing
Creativity
The Greatest Inventions of the Past 2,000 Years
The Next Fifty Years:
Science in the First Half of the Twenty-first Century
The New Humanists: Science at the Edge

As co-editor

How Things Are

CURIOUS MINDS

How a Child Becomes a Scientist

EDITED BY
John Brockman

VINTAGE

Published by Vintage 2005

6 8 10 9 7 5

First published in Great Britain in 2004 by Jonathan Cape

Vintage
Random House, 20 Vauxhall Bridge Road,
London SW1V 2SA

The Random House Group Limited Reg. No. 954009
www.randomhouse.co.uk/vintage

A CIP catalogue record for this book is available from the British Library

ISBN 9780099469469

The Random House Group Limited supports The Forest Stewardship Council® (FSC®), the leading international forest-certification organisation. Our books carrying the FSC label are printed on FSC®-certified paper. FSC is the only forest-certification scheme supported by the leading environmental organisations, including Greenpeace. Our paper procurement policy can be found at www.randomhouse.co.uk/environment

FSC www.fsc.org MIX Paper | Supporting responsible forestry FSC® C018179

Printed and bound in Great Britain by Clays Ltd, St Ives plc

Contents

Acknowledgments

I first began thinking about this book when, in Santa Fe on Christmas Day 2002, I spent much of the afternoon in a long, rambling conversation with Murray Gell-Mann, discussing his childhood.

The theme of *Curious Minds* came out of a conversation between Marc Hauser and Dan Dennett at a dinner in Cambridge, Massachusetts, several months later.

I wish to thank my U.S. publisher, Marty Asher, of Pantheon Books, and my U.K. publisher, Will Sulkin, of Jonathan Cape, for their encouragement.

I am also indebted to my agent, Max Brockman, who recognized the potential for this book, and to Sara Lippincott for her thoughtful and meticulous editing.

Introduction

JOHN BROCKMAN

In the spring of 2003, I attended a talk given by the mathematician Douglas Hofstadter and sponsored by Daniel C. Dennett's Center for Cognitive Studies at Tufts University. Later I gave a dinner in Cambridge whose guests included Hofstadter, Dennett, and the Harvard psychologist Marc Hauser, among other scientists. The wide-ranging and animated conversation that evening covered such topics as intuition, evolutionary biology, artificial intelligence, cognitive and neuroscience, music perception, and more. I recall thinking that dinner-table conversation doesn't get any better than this.

At one point, Marc Hauser turned to Dan Dennett and asked, "Can you remember when you got started thinking about these issues? How old were you? When did you get passionate about ideas?" Dan replied that when he was six an adult told him that since he was asking such interesting questions, he should become a philosopher. Doug Hofstadter said that from the first moment he could remember, he loved numbers and knew he wanted to do mathematics. For Marc, it wasn't until college that he discovered his specific interests. But what they all shared as children was curiosity and a deep passion for learning, whether specific or general. As one of the other dinner guests mused, "It all started when we were kids."

Curious Minds: How a Child Becomes a Scientist contains twenty-seven essays, contributed by some of the world's foremost third-culture scientists—public intellectuals who, in their writing, bridge the once formidable gap between science and the humanities. Their essays, from the mind and from the heart, are personal and accessible. No scientific background required!

As a starting point for the essays, I asked the contributors, "What happened when you were a kid that led you to pursue a life in science? What got you interested in your current field of research and inspired you to become the person you are? Who were your parents, your peers, your mentors? What were your turning points, your segues, your influences, your epiphanies, your accidents, pressures, conflicts, mistakes?"

My goal was to provide readers with insight into the early lives of some of the most interesting thinkers on the planet. And to motivate and inspire.

These essays convey a sense of the excitement that the authors evidently feel about the stated goals of the book. I must also note that that this group of scientists—a particularly unruly bunch in some cases—did not slavishly adhere to my mandate. Some of the contributors sail through their early childhoods and pick up their narratives in college and graduate school. But they are all worth knowing and listening to, and this book is a good place to begin.

The themes in the book are not predictable. Prepare to be surprised.

Richard Dawkins writes about the influence of reading Dr. Dolittle. David Buss learns about certain facts of life while working at a New Jersey truck stop. Janna Levin, in love with the cosmos, fantasizes about traveling to the edge of the universe. Nicholas Humphrey muses about the privileges that go with being born into a dynasty of scientists. Robert Sapolsky haunts the Bronx Zoo and wants to be a mountain gorilla. Steven Pinker turns the thesis of the book on its head. "Rather than childhood experiences causing us to be who we are, who we are causes our childhood experiences," he writes. Judith Rich Harris recounts her isolation, loneliness, and differentiation from others in her peer group.

Several contributors grew up surrounded by nature: Tim White writes about life in rural Southern California, and Murray Gell-Mann, as a child, regards New York City as a hemlock forest that had been overlogged.

Daniel Dennett, Steven Strogatz, Marc Hauser, and Doyne Farmer talk about the teachers and mentors who changed their lives forever.

Paul Davies and Lee Smolin recall their early encounters with members of the opposite sex.

The power and influence of books and the ideas associated with them is a theme of the essays by Alison Gopnik, Mihaly Csikszentmihalyi, Howard Gardner, and Sherry Turkle.

Mary Catherine Bateson, Freeman Dyson, and V. S. Ramachandran write about important parental influences.

Ideas and experiments informed the early lives of Lynn Margulis, Joseph LeDoux, Rodney Brooks, Jaron Lanier, and Ray Kurzweil.

Recently, I spent an afternoon in Santa Fe talking with Murray Gell-Mann, particle physicist, extraordinary polymath, and Nobel laureate. I have had long, rambling conversations with Murray over the past decade, mostly about scientific issues, ranging from quarks to complex adaptive systems. This time I was curious to learn about his childhood—about how he became interested in physics in the first place and about the evolution of his long and brilliant career. During our conversation, he told me a story.

In 1951, at the age of twenty-one, Murray was a postdoc at the Institute for Advanced Study in Princeton, where he often encountered Albert Einstein as both were on their way to work. "Albert Einstein was my father's hero," Murray said, "and I could have spoken to him, but at that time I didn't like the kind of people who introduced themselves and tried to strike up acquaintances with great figures." Murray happened to be in disagreement with Einstein's attempts to seek a unified field theory without incorporating quantum mechanics or the strong and weak nuclear forces. "If he had been working on something that looked promising," Murray said, "I would have felt that I had a legitimate reason to talk with him, but this wasn't the case. . . ." And now,

Murray told me, he was full of regrets that he had not simply asked the elderly Einstein about his thinking back in the early years of the twentieth century, "when he was carrying out the greatest physics research since Newton's. That would have been exciting!"

Perhaps as much as anyone else in today's culture, I find myself in an analogous privileged position: that of being able to talk with the world's leading scientists—and to cajole them into writing essays about their own early lives and the history of their own attitudes toward the world and toward what would become their life's work. Unlike Murray, who failed to engage Einstein in conversation in 1951, I have no intentions of letting this magnificent opportunity slip by!

CURIOUS MINDS

A Family Affair

NICHOLAS HUMPHREY

NICHOLAS HUMPHREY, School Professor at the London School of Economics and professor of psychology at the New School for Social Research, is a theoretical psychologist, internationally known for his work on the evolution of human intelligence and consciousness. His books include *Consciousness Regained, The Inner Eye, A History of the Mind, Leaps of Faith,* and *The Mind Made Flesh.*

On Boxing Day 1960, soon after breakfast, Gower Street in London was deserted. I and my grandfather, A. V. Hill, entered the anatomy department of University College through a side door and made our way stealthily upstairs to his laboratory. The atmosphere was morguelike, and a musty smell of formaldehyde hung in the air. Water dripped from the lab ceiling and splashed onto an umbrella raised over the bench. A clock ticked, oddly out of tempo with the dripping; otherwise there was an eerie stillness. Grandpa removed the lid from a basin filled with live frogs, picked one out, and eyed its strong thigh muscles. He put it aside in a glass jar and called me over to admire it. The dissecting instruments and pins were waiting beside the corkboard.

I was seventeen years old. I had been reading Hermann Hesse's novel *Steppenwolf,* and I thought of the Magic Theater, with the strange sign on its door: "Not for Everybody." I felt (not for the first time) that I had crossed a threshold into a place from which ordinary people were excluded. But in the novel the theater's door bore another sign beneath the first: "For Madmen Only." I was proud to be where I was, and in this company, but I was wary, too.

My grandfather had in fact chosen this day to go to work, when most normal people were still in bed sleeping off their Christmas dinners, for the sanest of reasons. Following on from the research for which he had won the 1922 Nobel Prize in Physiology or Medicine, he was now, at age seventy-five, conducting what he would later call his "last experiments in muscle mechanics." He had recently developed a much improved moving-coil galvanometer to measure the heat output during muscular contraction, but his new instrument was so sensitive to vibration that every car passing in the street outside, every footstep on the landing, created a false reading. So a day like this, which belonged only to him and me, was the ideal time to make a perfect measurement.

He could have done the experiment alone. But science for my grandfather was nothing if not a family affair, and he had long been in the habit of engaging his children and grandchildren as his assistants. This is his account of how he prepared for the Royal Institution Christmas Lectures in 1926:

> Of the suggestions for my Lectures, the best came from Janet, aged eight, who proposed that I should make experiments upon her. . . . The more I thought about it, the better it seemed. Fearful experiments I would make on all my children: Polly's heart should be shown beating; and her emotions should be exposed on a screen. David should be given electric shocks till sparks came out of his hands. . . . Janet should have the movements of her stomach (there is no decency in young ladies these days) shown to the audience on a screen. Then the noises made by Maurice's heart should be made to resound like a gun all round the lecture hall . . . and he would not be content till I had promised that he also should have electric shocks.

Now, a generation later, he had called on me to help him, as part of the tradition. Research assistant or sorcerer's apprentice? A bit of both.

At lunchtime we ate the cheese and cider that were Grandpa's standard fare. The cider, pale and dry, came from a press in the village of Ivybridge in Devon, where for many years he had owned a country house on the edge of Dartmoor. Prompted by the Devon associations and the intimacy of the occasion, he told me the story of how he had been able to date precisely the day he first set foot on the moor. He and his mother had been staying for the holidays on a farm nearby. Borrowing a gun from the farmer, he had gone out to shoot rabbits. Around midday, to his complete surprise, he saw a solar eclipse developing, with the sun beginning to be swallowed by the shadow of the moon. He took the glass from his pocket watch and smeared it with the blood of a rabbit he had just killed, so that he could watch the phenomenon in safety. Many years later he verified the date in an astronomical almanac: May 28, 1900, 2:30 P.M.

Grandpa never had much time for metaphysics ("the art of bamboozling people—methodically," he once told me). But on that morning he and I were developing an unusual bond, and now he let himself talk of things he would not normally have shared. There was a lesson in the story of the rabbit's blood and the eclipse. The sun, moon, and stars have one kind of destiny. Their times and courses are fixed by well-known laws. Newton could have predicted hundreds of years earlier exactly what would be seen at that place and time. But rabbits and boys—yes, and frogs, too—have another kind of destiny. It seems that we know neither the day nor the hour wherein fateful things will happen. What laws, if any, apply to human behavior?

Pavlov, whom Grandpa had counted as a friend and had several times visited in Leningrad, believed that there would one day be a science of the mind similar in rigor to the sciences of physics and chemistry. For that matter, so did Sigmund Freud, to whom my grandfather played host when Freud was made a Foreign Member of the Royal Society in 1938 and with whom he had got on surprisingly well. But what contrasting notions those two had of what science is! Grandpa had given me, some years earlier, a framed text of Pavlov's "Bequest to the Academic Youth of Russia"—or, as it became known, Pavlov's Last Testament—written just before his death in 1936 at the age of eighty-seven. This is the passage he marked out for me:

> Never attempt to screen an insufficiency of knowledge even by the most audacious surmise and hypothesis. Howsoever this soap-bubble will rejoice your eyes by its play, it inevitably will burst and you will have nothing except shame. . . . Perfect as is the wing of a bird, it never could raise the bird up without resting on air. Facts are the air of a scientist. Without them you never can fly. Without them your "theories" are vain efforts.

Grandpa loved that image of the soap bubble. Just right, or so he thought, for describing Freudian theory. Later that day, when we returned to his study, he pulled out an essay written in 1925 by his brother-in-law, John Maynard Keynes:

> I venture to say that at the present stage the argument in favour of Freudian theories would be very little weakened if it were admitted that every case published hitherto had been wholly invented by Professor Freud in order to illustrate his ideas and to make them more vivid to the minds of his readers. That is to say, the case for considering them seriously mainly depends at present on the appeal which they make to our own intuitions as containing something new and true about the way in which human psychology works, and very little indeed upon the so-called inductive verifications, so far as the latter have been published

> up to date. . . . [Freud] deserves exceptionally serious and entirely unpartisan consideration, if only because he does seem to present himself to us, whether we like him or not, as one of the great disturbing, innovating geniuses of our age, that is to say as a sort of devil.

Huh! Didn't *that* put Freud nicely in his place!

I listened and watched and took things in. I moistened the frog's muscle with Ringer solution. I had just left school, and the plan was for me to go to Cambridge the following October, with a scholarship to read math and physics. I knew next to nothing about biology. But my grandfather had other ideas for me. He himself had started out as a mathematician, only later to discover the world of biophysics. Now, he implied, the next real challenge lay in the behavioral sciences. A few weeks later, he arranged for me to spend six months at the Marine Biological Laboratory at Plymouth as a lab assistant to his protégé, Eric Denton, where I could learn—at any rate, begin to learn—about life.

And so I went, and so I did.

The poet W. H. Auden wrote: "When I find myself in the company of scientists, I feel like a shabby curate who has strayed by mistake into a room full of dukes." Possibly none of us except a duke can know what it feels like to be born to be a duke. Quite special, I imagine: One would have a sense of intrinsic superiority, of rights of access and freedoms from restraint not allowed to ordinary people. But I do know as well as anybody what it feels like to be born into a dynasty of scientists. Quite special, I can confirm, and somewhat the same.

A. V. Hill, my mother's father, was a scientist in the grand mold: Nobel laureate, member of Parliament for Cambridge and Oxford Universities in the Churchill war administration, champion of intellectual freedoms and responsibility around the world.

He played a crucial part in arranging the flight of Jewish scientists from Hitler in the years before the war. Throughout my childhood, at my grandparents' house in Highgate, there were always visitors with heavy mid-European accents and twinkling smiles, in excited discussion of new discoveries—who would receive, as the years passed, Nobel Prizes of their own.

My great-uncle Maynard Keynes died when I was two, but his intellectual presence hung over our family, and his wife, the Russian ballerina Lydia Lopokova, with all her Bloomsbury connections, lived on as a spritely babushka. Maynard's brother, Geoffrey, was a surgeon and medical historian; *his* wife, Margaret, was a granddaughter of Charles Darwin.

My mother, Janet (she of the moving stomach), became a doctor and later a psychoanalyst who worked with Anna Freud. Of her brothers and sisters, Maurice (of the resounding heart) became a geophysicist whose work was central to establishing the reality of continental drift. David (of the sparking hands) became a biophysicist, who, like his father, did research on muscle. Polly (of the exposed emotions) became one of the first economic anthropologists and studied the workings of the West African cocoa trade. Both Maurice and David were Fellows of the Royal Society, an institution that Grandmother Hill—with six of her immediate family as Fellows—came to regard as her private club.

My father, John Humphrey, was an immunologist and director of the National Institute for Medical Research, where he did seminal research on antibody formation. But like others in the family he was also deeply engaged in social and political issues. He was the founder of the Medical Campaign Against Nuclear Weapons, whose offspring, the U.S.-based International Physicians for the Prevention of Nuclear War, would later win the Nobel Peace Prize. His father, Herbert Humphrey, was an engineer and inventor. We had at home a *Vanity Fair* portrait of Grandfather Humphrey, captioned (after one of his inventions) "The Humphrey Pump"; but the invention of which, as a boy, I was secretly prouder was a one-man "manned torpedo" he designed in the First World War for use against German ships and for which he proposed himself as the first pilot—an offer that Churchill, then at the Admiralty, declined.

My grandfather's brother, Willie, had been a brilliant mathematician at Cambridge. But he turned to the church and became a missionary in West Africa, where he ran into trouble with the natives and was beheaded (and, so we always imagined, eaten). I never knew him, but I was given the telescope with which he used to watch for the mailboat coming into Freetown harbor—and along with the telescope a pathetic telegram, sent to his sister in England by a friend after his murder, saying simply: "Willie—Gravest News—More Follows." This sister, my great-aunt Edith, had wanted to be scientist, too, but there were no openings for women at British universities in the 1890s, so she went as a doctoral student to Zurich, where she attended lectures by the great Russian chemist Dmitry Mendeleyev, inventor of the periodic table. Later she became England's first female industrial chemist. She lived till the age of 103 and was a regular presence at our dinner table.

My immediate family was a large one. At home there were seldom less than ten at any meal, and during school holidays generally more. We lived in a huge house—Scottish baronial in style, with twenty-six rooms and over an acre of garden—in Mill Hill, North London, close to my father's research institute. I had four brothers and sisters and two orphaned cousins living at home, and another fifteen cousins within easy reach. We children went round in droves, stayed with one another in nearly unmanageable numbers, and met up regularly at my grandmother Hill's Sunday tea parties.

Stephen Hawking, then sixteen, came to live with us for a year while his parents were in India in 1958. Stephen was an intense, rather quizzical schoolboy, whom I remember (from my position two years his junior) as somewhat bossy. When I saw him many years later at his fiftieth birthday party in 1992, leading the dancing in his wheelchair, I reminded him of his efforts to teach my family to dance Highland reels (but I forbore to remind him of my more salient memory of him, marching up and down the hall of our house wielding a swagger stick and addressing an imaginary platoon of hapless schoolmates).

As children, we lived and breathed science, though of course we didn't know this at the time. Our sprawling basement rooms were full of apparatus: prototype engines of my grandfather's, pumps and torpedoes, lathes and jigsaws, Meccano sets, photographic apparatus, Wimshurst electrical machines, microscopes, aquariums. We spent Saturdays running round the corridors of my father's institute. We went on outings to my uncle Maurice's observatory in Cambridge. We went on trips on the research ships out of the Marine Biology Laboratory. We accompanied Stephen's family on expeditions in search of flint arrowheads in the woods at South Mimms.

But the major event of each week was the visit to my maternal grandparents, the Hills. The company at these Sunday parties usually spanned three and sometimes four generations, with my grandfather's colleagues and students invited to sit down with his offsprings' offspring—highchairs on one side, wheelchairs sometimes on the other. After a formal tea of sandwiches and cakes, the grown-ups would thankfully retire to the drawing room to talk science and politics, while the children were turned out into the garden to amuse themselves (or rather the several gardens, for my grandmother Hill, who liked to have a lot of everything, had systematically bought up all the neighboring properties).

However, my grandfather would not neglect us for long. Almost every week he devised some new game or experiment: frog races, archery, kite-flying, or perhaps, if the weather was bad, a magic lantern show. On one memorable occasion, he produced a sheep's head acquired from the butcher, and placing it on the kitchen table (to the cook's great distress), he dissected it in front of us. He carefully took apart one of the eyes and held the lens up for us to look through. I gazed through this jewel, out of the kitchen window over the garden, the lawn, the swing, my grandmother's prize dahlias: a beautifully clear image and *everything turned upside down.*

I still wonder about the effects of having grown up in such a strangely privileged environment—of having been bounced, as it were, into the world of science. Fifty years on, it is easy to make too much of it. Each of us is who we are, and we must each have had *some* sort of childhood. Who's to say whether any particular factor carried the weight that our self-narrative now likes to attribute to it.

The "nurture assumption" has been under attack in recent years. Yet I am inclined to think that in this case the causal pattern is real and undeniable. What I gained from this childhood environment was a sense of intellectual entitlement—a right to ask questions, to pry, to provoke, to go where I pleased in pursuit of knowledge. As a boy I was always taken by the words on the front page of my British passport: the demand that the bearer should "pass freely without let or hindrance." I grew up feeling that I carried a similar warrant to explore anything I chose, that I could indeed safely cross into areas "Not for Everybody."

To be a good scientist surely requires just such audacity. How else dare anyone do what a scientist is required to do: to challenge Nature to undress before one's eyes? One might claim an interest in Nature's secrets on several different grounds, but nothing compares, I suspect, to the feeling that one has some kind of ancestral droit du seigneur.

And yet there is, I'm almost afraid to admit, a downside, too. Indeed, I have come to realize that to take these rights of access for granted may not be entirely a good thing. In my own case, the problem I now see is that I never had to *struggle* to become a scientist and never experienced any real surprise or sense of achievement at having made it. And I confess that because of this, the rights and duties of the role have not always weighed with me as seriously as they ought to have. In particular, I have never known the proper worry that if I did not watch my back—finish one project before starting another, respect academic boundaries, get the right research grants, sit on the right committees—then my privileges might be withdrawn.

Pavlov's Last Testament ends with a warning: "Remember that science demands from a man all his life. If you had two lives, that would be not enough for you. Be passionate in your work and your searchings." My grandfathers on both sides, like Pavlov himself, were the first of their line to enter the world of science. The passion they put into their work was the passion of scientists who daily counted their blessings for being allowed to do science—and who were determined to repay the debt with single-minded dedication. Although A. V. Hill did much else besides, his first and his last experiments were on the thermodynamics of muscular contraction.

Two generations down the line, as I look back on the somewhat random walk of my own research career—in neuropsychology, ethology, evolutionary psychology, philosophy of mind—cherry-picking the most exciting problems as I went along, I marvel at the charmed life I have had so far. But I wonder whether, in the end, having been born to be a scientist has not undercut my right to call myself a scientist at all.

The Bungling Apprentice

DAVID M. BUSS

DAVID M. BUSS is a professor of evolutionary psychology at the University of Texas at Austin. His books include *The Evolution of Desire: Strategies of Human Mating* and, most recently, the second edition of *Evolutionary Psychology: The New Science of the Mind.*

Mating is the subject that has dominated my adult scientific life. Attempts to trace the causal paths that led me to this unusual topic are necessarily speculative. The stories we tell about our past, the selective inclusions, omissions, and shadings, undoubtedly get slanted toward ends other than unbiased description. It's unlikely that our accounts will be entirely unaffected by considerations of status and reputation. Nonetheless, there may be value in the stories regardless of the spin we put on them. One theme in mine is eloquently captured by the novelist Vladimir Nabokov, looking back on his youth:

> The ecstatic love of a young writer for the old writer he will be someday is ambition in its most laudable form. This love is not reciprocated by the older man in his larger library, for even if he does recall with regret a naked palate and rheumless eye, he has nothing but an impatient shrug for the bungling apprentice of his youth.

It's the naked palate and the bungling that most characterize my early days.

My family valued education, but my academic career did not take hold at first. When we were growing up in Indianapolis, my older brother was thought of as the brilliant one, destined for greatness as a scientist. With his prowess on the high school chess, math, and debate teams, his straight As, and his perfect score on the quantitative SAT, there was no reason to doubt it. My younger sister was the creative one, destined for greatness in a field as yet unspecified. I, on the other hand, though a mediocre academic performer, had good eyesight; perhaps, my father suggested, I could become an airline pilot. My grades sank further in junior high school and leveled off at roughly a C+ in high school. My class attendance became sporadic. Recreational drugs beckoned. I decided that school had nothing to offer. The details of my adolescent immaturity, which I rationalized as rebelliousness, are best forgotten. One school counselor summed it up, shaking his head with disappointment: "Buss, you need guidance."

And so I did, but it was not forthcoming immediately. After two drug-related arrests (the charges were subsequently dropped), I decided to leave school. I got a job at the first place I applied, a truck stop outside New Brunswick, New Jersey. The interview consisted of one question: "Are you willing to work nights?" I said I was. So I got the twelve-hour graveyard shift. From 7:00 P.M. to 7:00 A.M., I pumped gas, bumped tires, checked oil, and performed other tasks of comparable complexity. My supervisor at the truck stop was a forty-year-old black man whom we called Sergeant Tony, presumably because of his previous army experience. During slow nights, Sergeant Tony and I talked about life, and sometimes the conversations turned to women. One evening, as I was spouting the peace-and-love values prevalent in the early 1970s, Sergeant Tony patiently expressed his dismay at my naïveté: "David, the man always pays."

"But Sergeant Tony, what about free love?"

He just shook his head. "The man always pays," he said. I refused to believe such nonsense—the sexes were supposed to be equal, love and sex were freely exchanged, and I assured him that the world was moving toward those liberated views. Fifteen years later, my research on human mating in thirty-seven cultures located on six continents and five islands has shown that Sergeant Tony was not too far off.

I also recall proclaiming, in between servicing trucks, that jealousy was an immature emotion that only hung-up, uptight, unliberated people experienced. I announced that my girlfriend's body was her own, that she could have sex with anyone she chose and I would not be bothered in the slightest. I did not have a girlfriend at the time; a year later, when I did, I found that my feelings about the matter had changed. In my professional research on jealousy, which has confirmed the male obsession with the sexual exclusivity of a partner, I discovered a finding that coincided perfectly with that previously forgotten truck-stop conversation: that men without sexual experience report less sexual jealousy about the hypothetical infidelity of a romantic partner than do men who have had actual sexual experience.

Most of the action happened at the back of the truck stop, where some of the workers lived, out of sight of the owners, who occasionally dropped by to make sure that work was proceeding without problems. The staple beverage was cheap beer. Workers, the unemployed, hangers-on, and even Sergeant Tony would be visited by the occasional woman for brief sexual trysts. Although I never witnessed it, I'm sure cash changed hands. A cocaine dealer in a black hat and white Cadillac made sporadic appearances, but few had money to spend on expensive drugs. Mostly, truck-stop life consisted of people trying to get along, get through the night, and grasp whatever straws of satisfaction were afforded out back. Despite the minimum wage most of them earned, I don't think they were any less happy than anyone else. After three months at the truck stop, I felt I had learned more than I had in over a decade of attending public schools. And perhaps I had, but the learning experience came at a cost. One night a drunken truck driver threatened to "take a tomahawk to your long hair." Another night I was clubbed by another young man who was just trying to prove how tough he was. I decided there must be better ways to earn a living.

Two chance events turned my life around. The first was meeting a woman on an airplane en route to Amsterdam, where I had fantasies of fleeing to escape America entirely. She was a little older than I was and had a master's degree in biochemistry; I had just finished high school, taking night classes. She had been raised in Bremen, Germany; I was from Indianapolis. But none of this proved a barrier. We moved in together.

The second chance event was being accepted in 1971 to the University of Texas at Austin through a trial lottery system—abandoned the following year—in which those not in the top ten percent of their graduating class were selected randomly. At Austin, for the first time in my life, I was intellectually captivated by my classes. I encountered the theory of evolution through courses in geology (the evolution of life) and astronomy (stellar evolution). I had not known that there were theories attempting to explain how everything in the universe, including all life-forms on Earth, had originated and arrived at their current forms. I wanted to know more. Deep time captivated my imagination.

The two chance events dovetailed. Each night when my girlfriend returned home from her job at a genetics lab and I returned home from the university, she made me tell her everything I had learned that day. She loved learning and was envious because I got to take classes while she "counted fruit flies," as she put it. I learned to talk, and as my older brother is fond of saying, I haven't shut up since.

In my junior year, I knew that I wanted to become a scientist and that the human mind was the territory I wanted to explore. I took a course with David Hovland, son of the famous Yale psychologist Carl Hovland. In writing a term paper for the course, I decided to go out on a limb. I titled it "Dominance/Access to Women," and in it I proposed that men had evolved a powerful status-striving motivation and the sole reason was that dominant men gained sexual access to women. This was my first fumbling attempt at forming an evolutionary hypothesis. I brought in primate evidence from savanna baboons. I introduced data from an ethnography about the polygynous Tiwi tribe, whose headman had twenty-nine wives. And I argued that the same dynamic was playing itself out in modern America. Dr. Hovland asked me to present the paper to the class. To my surprise, it was greeted enthusiastically. A year later, when I began graduate school at the University of California at Berkeley, I conducted my first empirical study on human mating using evolutionary theory.

At the time, there was no field of evolutionary psychology, and evolutionary theory was virtually absent from all the social sciences. But I pursued it on my own, despite the indifference to evolution expressed by most of the psychology professors and the puzzled, sometimes derisive, comments from my fellow graduate students. By the time I received my PhD in 1981 and accepted a job as an assistant professor at Harvard, I knew there was no turning back from evolutionary theory—the field of psychology just hadn't realized it yet.

Are these traces of my past sufficient to explain my preoccupation with the evolutionary psychology of mating? Perhaps a few more details would help. From an early age, I found myself fascinated with females. At seven or eight, I became irresistibly drawn to the girl next door. I had no name for my feelings, but later I was sure it was love. When she freed me during a neighborhood game of tag, in which those tagged had to remain "frozen" until being released by a peck on the cheek, I felt a hot, pleasurable burning sensation on my face that lasted for hours.

Did these childhood experiences somehow create some causal vector that motivated me to focus on mating in my professional life? Possibly, yet I doubt that my experiences are unique—although all this is speculation, since little is known scientifically about childhood attractions. As I grew up, I found out that nearly every one of my peers was mesmerized by mating. School gossip revolved around it: Attractions, repulsions, mate competition, mate poaching, mate switching, and sexual conflict permeated our social life beginning in sixth or seventh grade and possibly earlier. Mating obsessions continued to suffuse the social lives of my friends and acquaintances in high school, college, graduate school, and beyond. I was certainly not alone in my fascination with it.

Once I became enchanted by evolutionary theory, however, mating became a natural. Differential reproductive success is the engine of evolution; nothing is closer to reproduction than mating. If there existed one domain in which Darwinian selection would be expected to sculpt finely honed adaptations, mating would be it. Even survival is secondary: It is now known that there are adaptations that actually decrease the chances of survival but have evolved because they increase success at mating. Given the intensity of sexual selection, the acceleration of the evolutionary process through antagonistic coevolutionary mating "arms races," and the sheer number and complexity of mating problems that human beings must solve, there may be no suite of human psychological adaptations more complex, sophisticated, and mysterious than those of human mating.

When I began to scour the psychological literature for theories and research on mating, however, I found a virtual vacuum. Aside from an occasional study of attraction, practically nothing was known scientifically about how people compete for mates, derogate their mating rivals, emit attraction signals, preferentially choose mates, guard mates, and retain some mates while jettisoning others. Thus I was in a unique position to explore a domain of supreme theoretical importance—a domain personally captivating but about which almost nothing scientific was known. The confluence of all these factors has undoubtedly dictated my choice of a scientific path. My colleagues sometimes tell me that I'm fortunate that my professional life is devoted to such an interesting suite of topics. They're right.

Mountain Gorilla and Yeshiva Boy

ROBERT M. SAPOLSKY

ROBERT M. SAPOLSKY is a professor of biological sciences at Stanford University and of neurology at Stanford's School of Medicine. His latest book, *A Primate's Memoir,* grew out of his annual trips to East Africa to study a population of wild baboons. He is also the author of *The Trouble with Testosterone and Other Essays on the Biology of the Human Predicament* and *Why Zebras Don't Get Ulcers: An Updated Guide to Stress, Stress-Related Diseases, and Coping.*

How did I wind up as scientist? By all logic, I should start with *Gilligan's Island,* a sitcom that entranced me when I was an eight-year-old growing up in Brooklyn. In it, an unlikely collection of seven people go on an afternoon's boat ride out of Hawaii, get waylaid by a storm, and wind up stranded on a desert island, where they remain for years. The motley crew includes the skipper, his first mate, a wealthy upper-crust couple, a famous actress, a farm girl, and "the Professor," who is otherwise nameless. He has every book ever written somewhere in the trunk he was marooned with; he can answer any challenging question you can think of; he is forever saving everyone by rigging up some sort of scientific device. The Professor can do anything (except get them off the island, of course).

While all this was impressive, what really got to me was his presumed connection to Mary Ann, the pretty farm girl in flannel shirt and pigtails. This connection I derived solely from the show's theme song, which went "There's Gilligan, the skipper, too, a millionaire and his wife, a movie star, the *Professor and Mary Ann. . . .*" Because their names were linked, I assumed that the two of them must have had something going. In my prepubescent fog, this involved a lot of hand-holding. So it was only natural that I wanted to grow up and be the Professor and spend my time out in some remote field site.

That should explain everything, but it doesn't. There are still two questions to answer: Why have I wound up studying primates and the brain? And what are the sources of the emotional baggage I bring to that work?

I'm half neurobiologist, half primatologist. For a large part of the year I'm the former. In my laboratory at Stanford we study the interactions between stress and neurological disease. We try to understand how stress hormones can damage the nervous system—that is, make neurons less likely to survive neurological insults—and we design gene-therapy strategies to try to save neurons from those effects. So far, we haven't been all that successful. For the remainder of each year, I study a troop of wild baboons in the Serengeti in East Africa. I've been doing that for twenty-five years, going back to the same animals. The work there is the flip side of the lab work: Given all the bad things that stress can do to you, why is it that some of us are better at handling stress than others? When I'm with the baboons, I study what their social rank, personality, and patterns of social affiliation have to do with who among them falls victim to stress.

The primatology part came first. When I was eight or so, I decided I wanted to study apes in the wild. It really wasn't a particularly coherent, cognitively shaped interest, just an outgrowth of the earlier dinosaur stage. I had started with the dinosaurs, but my father was an architectural historian who had done some archaeology in his time, and that got me to the King Tut's tomb stage. From there, I progressed to the bones of our hominid ancestors. But at some point I started going to the Bronx Zoo and the American Museum of Natural History, and the primate exhibits simply did something to me that the bones and potsherds couldn't approach.

It wasn't just the obvious—that dynamic, living primates can be more interesting than trying to imagine a primate ancestor from a fragment of skull. Something resonated, in a way that I still feel but can't explain. I was a fairly solitary, misanthropic kid, probably atypically likely to get caught up in some obsession, but the intensity of the response still puzzles me. It wasn't just that primates seemed fascinating; they seemed comforting in some primal way. It wasn't that I wanted to go off and live with, say, mountain gorillas: I wanted to *be* one. Primates grabbed me in a way that still makes me ache when I see them.

I became obsessed. I had primate pictures up all over the place, lived for Time-Life books on the subject, made vague inroads into the academic literature. I started learning Swahili in high school in preparation for doing fieldwork in Africa. I wrote fan letters to primatologists, a few of whom, now emeriti, remember them as being pretty unmodulated and intense.

It was an intellectually unfocused vocation. I was going to go into the field and combat poachers to save primates from extinction. I was going to do lab primatology and find a cure for cancer. I was going to make discoveries about primate social behavior that would bring about world peace. I was going to prove that nonhuman apes had secretly evolved language and religion, and that yetis were for real. All this was the stuff of *Gilligan's Island.* It finally took a more coherent form in college. I entered Harvard and took up biological anthropology in the mid-1970s, just as one of the stars of Harvard's biology department, E. O. Wilson, was publishing *Sociobiology,* provoking one of the great academic shitstorms of recent decades. The Harvard primatologists were all in Wilson's camp. *Sociobiology*'s thesis was that all animal behavior (humans included), even altruism, should be seen as the product of evolution and Darwinian fitness.

But something about Wilson's focus bothered me. I was probably uncomfortable with the controversy. *Sociobiology* had a right-wing tinge to it that was more imagined than substantive, and the left was ferociously critical of it, accusing Wilson of, among other things, justifying the white male patriarchy as natural and biologically ordained. My visceral sympathies and personal style were much more in sync with the leftist critics, while it seemed to me that most of the leading sociobiologists around Harvard were southerners who smoked and drank and worked out their ideas while playing poker all night. They basically scared the bejesus out of me.

The other problem I had with the purely sociobiological take on things was how shrill and one-dimensional it seemed, at least then. There were some rabidly hostile debates going on in academia at the time, swirling around sociobiology, behaviorism, genetics, and IQ, and most of the major heavyweights in these fights—besides Wilson they included Richard Lewontin, B. F. Skinner, Noam Chomsky, and Stephen Jay Gould—were in Cambridge. Everyone was arguing with each other; the fireworks were amazingly entertaining, and the contentious atmosphere did away with subtlety, boxing people into some pretty strident corners. Everyone had become Isaiah Berlin hedgehogs:

"All of behavior is rooted in evolutionary fitness."

"People who think of human behavior as selectively evolved have hidden agendas."

"Environment is of paramount importance in shaping who we are."

"IQ is largely heritable."

And so on.

I found a kind of refuge with my academic adviser, the person who has probably had a greater intellectual influence on me than anyone else, a then assistant professor named Mel Konner. He had a pronounced distaste for the one-sidedness of all the debaters and instead thought in a wonderfully subtle, multidisciplinary way. Evolution and ecology shape who we are, certainly. But we need to consider neuroscience and endocrinology, developmental psychology and philosophy, with literature thrown in for good measure. None of the major mugwumps going at one another paid much attention to him, simply because he was almost pathologically nuanced in his thinking, unwilling to say anything doctrinaire enough to trigger another round of mud-wrestling.

Something about this synergistic approach clicked with me, seemed tremendously satisfying and right, and has been with me ever since. Thus I do research in a hodgepodge of disciplines. In the field, I'm part animal behaviorist, part sociobiologist, part endocrinologist/physiologist, and sometimes I find myself having to make sense of the work of sociologists or economists. And in the lab, I study a mixture of endocrinology, neuroscience, and molecular biology, with a lot of clinical science on the side.

When it works well, there's a useful cross-fertilization. When it doesn't, I can feel like a dilettante skating on thin ice. And it complicates my sense of my own identity as a scientist. Am I a field biologist in hiking shoes or a lab scientist? Do I protect animals or do I kill them? Am I most interested in basic science or in doing something about human disease? Do I hang with baboons in Africa because they provide good data or because I love them?

None of these questions have clear answers for me.

And now to the emotional source of what I do. My emotional life is very tied up with my science. I really get worked up about science—what others have discovered, what I am discovering, what I am not able to figure out. I desperately want to make scientific breakthroughs that are going to help people. And I see science as an imperative, a weapon against antiprogressive forces in society, against right-wing yahoos, against religious intolerance.

I think about that aspect of science a lot, write about it, lecture about it—ad nauseam, no doubt. This adamant secularism arose in adolescence. I was raised in an Orthodox Jewish household—separate meat and dairy sinks, utensils soiled in violation of the kosher laws buried in soil in the flowerpots, the whole thing. For most of my childhood, I was really into it. Somehow I saw no contradiction between this and my fascination with primates. Evolution and one's religion were simply in different compartments.

You could be a mountain gorilla and a yeshiva boy at the same time. Religion was basically nothing more than being obedient and ritualistic for its own sake.

The difficulty started at age thirteen. At Passover, for the first time, I found that I was having tremendous problems with the internal contradictions of the Exodus story. Not the usual: Why did the horses have to drown? What had the firstborn done to deserve death? Instead, I was wrestling with the issue of free will and volition. Moses says to Pharaoh, "Let my people go." Pharaoh refuses. A plague ensues. Pharaoh says, "I give up. Go!"

"And then God hardened Pharaoh's heart."

So Pharaoh is compelled to reverse himself and as a result is punished again. Where's the responsibility, if God can mess with Pharaoh's metaphorical heart? And then why should Pharaoh—not to mention the cows of Egypt—suffer the divine wrath? This was bothering me. I was also struggling with some commentary on the Talmud stating that when bad things happened to people, it was because they had done something to deserve it—an appalling lesson to teach to folks just coming off the Holocaust.

But what really got me was my discovery one day of an ancient rule from the time of the temple in Jerusalem. According to the text our rabbi was teaching us, a man couldn't be a priest if he was a dwarf or a cripple.

"What's that about?" I asked. (At the time, I wore leg braces, because of a bone disease, and as a result regularly had the crap beaten out of me by the tough kids in my neighborhood.)

"Well, isn't it obvious?" the rabbi answered. "It would be an insult to God to have His temple presided over by someone blemished like that."

I was thunderstruck. God decrees who is a dwarf or a cripple. How can God be insulted by something He intentionally created? How can God punish the cripple for what God Himself has wrought? What's the deal with this? I was in a terribly confused, agitated state. By evening that agitation had given way to a sense of betrayal and a fury the likes of which I'd never experienced in my life. Finally, spent with rage, I followed the down-the-rabbit-hole illogic to its logical conclusion, which was to say my evening prayers, exalting the god I'd just been railing against. But internal contradiction couldn't persist. Two nights later, I woke up in the middle of the night with the sudden cold realization: *There is no God. This is gibberish.*

Since then, I've had no religion, in fact no capacity for spirituality of any sort whatsoever. There is no facet of my life—love, parenting, mulling over why we are here—that I view outside the context of mechanistic science. For me, there is no Divine Watchmaker—no top-down volition, no purpose, no cause beyond what emerges from the complexity of biological systems. This is not a cold point of view: I am as intensely emotional now as I was at the age of thirteen, and I don't find science and emotionality to be at all contradictory. Nor do I believe that science is an emotional substitute for religion. But for me, it has finally made the religious worldview impossible.

the family spent [illegible] ages [illegible] this was [illegible] the [illegible] Boat [illegible] Regent [illegible] [illegible] the god of the [illegible] against the [illegible] [illegible] That night [illegible] of the [illegible] There is no God. I'm fairly certain.

Not that I've had no [illegible] at any [illegible] whatsoever. There is no [illegible], including [illegible] mechanistic science. [illegible] no purpose, no cause [illegible] complexity of biological systems. [illegible] of the age and [illegible] that [illegible] science [illegible] I believe that [illegible] unlikely [illegible].

Safety in Numbers

MIHALY CSIKSZENTMIHALYI

MIHALY CSIKSZENTMIHALYI is the former chairman of the psychology department at the University of Chicago and currently Davidson Professor of Management at the Claremont Graduate University in Claremont, California. His books include the best-selling *Flow: The Psychology of Optimal Experience, The Evolving Self, Creativity,* and *Finding Flow.*

I was born in Italy, of Hungarian parents. Currently I am a U.S. citizen, but in order to own property near where I was born, I also need a Croatian passport. If this sounds confusing to you, imagine how it must be for me. Yet having been a "displaced person" for so long has its advantages. Without a comfortable cultural script for interpreting life, one is left to devise one's own. For better and for worse, my research and writing have been heavily influenced by the lack of roots.

I remember some years ago how thrilled I was when I read the first pages of Descartes's *Discourse on Method.* He goes to some lengths to explain that although he had been taught in the best schools in France, by the best teachers, he never really learned anything he was sure of until he started traveling and realized that what was true in France was not necessarily true in Germany, and vice versa—and that the Dutch regarded the intellectual accomplishments of both countries with contempt. This realization forced him to come to terms with first principles—with knowledge that was not based on hearsay but was rationally derived from the uncluttered workings of the mind. (It is interesting to reflect on the difference between Descartes's reaction to the relativity of knowledge and that of many modern social scientists who, disheartened by the despair that enveloped the West after the First World War, gave up on the notion of a universally valid rationality. Descartes saw parochial knowledge as a hindrance to understanding, while the cultural relativists see it as the only understanding there is.)

Reading Descartes was a revelation that helped me understand my own strivings. But it was not through delving into such issues that my professional interests took shape. One of the early "crystallizing moments," to use the cognitive scientist Howard Gardner's phrase, occurred in July 1948, when I was living in Rome and about to turn fourteen years old. It was shortly after a right-wing fanatic had gunned down and almost killed Palmiro Togliatti, the leader of the Italian Communist Party. The attack took place close to election time, and the question everyone debated was whether Italy's democracy could survive this crisis or if extremists from both sides would take to the streets and a new dictatorship would emerge to restore order, as had happened twenty-six years earlier, when Mussolini was allowed to seize the reins of government in a similar predicament.

But during the teenage years, politics takes on a much simpler, more familiar face. In my case, the incident became an excuse for good-natured taunting between my friend Silvio and myself. One lazy summer afternoon, I told him he'd better watch out, because if the Communists took over he would be in trouble. After all, the neighborhood he lived in was full of Reds and they knew he had two uncles who were priests. Silvio answered suavely that it was I who would be in trouble if the Communists took over, because my neighborhood was more Red than his and they probably knew that my father did not go back to Hungary because he didn't like the Russians who had taken over the country. Somehow this friendly sparring degenerated into a heated argument about who had better watch out more and whose neighborhood was the deeper Red. We started with insults and verged on fisticuffs when I had a brilliant idea about how to save our friendship: "Wait," I said. "Do you think if a newsstand sells more copies of *L'Unità* and *Avanti* than it sells of *Il Tempo* and *Il Messagero*, then the neighborhood is more likely to be Communist?" Silvio agreed that if the official papers of the Communist and Socialist Parties were in greater demand than those of the Christian Democrats, it would be a reasonable indication that the inhabitants of the surrounding area were politically on the left. Because World War II was still fresh in Italian experience, few people were trusting enough to pay up front for a newspaper subscription; they preferred to buy their newspapers at the corner kiosk, so newsstand sales should adequately represent the reading habits of the population.

Having agreed on first principles, we canvassed his neighborhood for newsstands, then did the same in mine. In the following days, we stationed ourselves on the corner in front of first one, then another kiosk and kept a record of the papers snatched up by the customers as they hurried by. At the end of a dozen days or so, we had covered all the kiosks, had thousands of check marks under the headings of the major dailies, and were ready to make sense of our efforts.

Yes, but how? We had no idea that there existed branches of the social sciences (what were they, anyway?) that had developed statistical methods for evaluating data such as we had collected. We had no idea that such things as surveys, polls, or market research existed. We just wanted to prove the other guy wrong. At first I exulted: Clearly Silvio's kiosks had a lot more check marks in the Red dailies' columns than the kiosks in my neighborhood. So I had been right! "Not so fast," said Silvio. "I also have more check marks for the papers of the Democrazia Cristiana. It's not the total number that counts, but the relative frequency."

"How could that be? If there are more Reds in your neighborhood than in mine, then that can only mean that yours is a more Communist environment."

"It does not, because the number of Communists in my environment is diluted by the even larger number of non-Communists!"

Thus in defending ourselves we had reinvented a statistical wheel, discovering some of the principles on which research is based, and now we understood in our very guts how easily one can cheat with numbers. But the outcome of my first exposure to empirical research was not cynicism. Instead I was elated by the possibility that the conflicting claims of ideological or self-serving arguments could actually be tested against relevant evidence. The evidence might leave a lot to be desired, and the results might not bring satisfying clarity, but some such process seemed better than the hocus-pocus most adults relied on to bolster their arguments.

A few years earlier, as the war in Europe was ending, I had had the advantage of observing closely some important grown-ups. They were generals, ministers, judges, heads of government offices—movers and shakers of the day. In normal circumstances, a ten-year-old boy would have had no chance to see them in action with their hair down, so to speak. But the war had erased many of the usual boundaries between public and private, old and young, weak and powerful. We brushed shoulders in air-raid shelters, offices hastily reassembled in hotel bedrooms, in trains, in parks. Like wasps disturbed in their hive, the elite of the ancien régime were bumbling about in the vain hope of regaining their cozy way of life. The overwhelming impression I derived from this experience was that most grown-ups had no clue as to what was really going on. Neither education, power, income, nor renown seemed to help people understand what was happening under their very noses. If Allied troops crossed the Rhine, some well-informed person was sure to come up with a verse from the *Prophecies* of Nostradamus that clearly predicted, four centuries before the event, that in 1944 Anglo-Saxon invaders would suffer a conclusive defeat once they crossed a line joining the cities of Metz and Essen. If someone fretted that Russian troops were getting dangerously close to central Europe, someone else could be counted on to list the reasons why the Russians—basically an uneducated race of serfs—were too dumb to stand up to the Wehrmacht. The most bizarre theories were advanced with the utmost seriousness, and gossip raged unchecked, making a mockery of even the concept of truth. Compared with that epistemological witches' brew, the game Silvio and I would play a few years later was a ray of hope. How much niftier it was to actually know what you were talking about! I wasn't aware of it then, but I was hooked on using empirical methods to understand human experience.

The road from that early epiphany to a PhD in psychology was not a straight one, however. A second stage began when I was about seventeen years old, after I heard Carl Jung give a talk on flying saucers. This was during a skiing holiday in Switzerland that turned out to be too late in the season, since the snow on most of the slopes had melted. There was a simple reason for attending the lecture: I didn't have enough money to go to the movies and the lecture was free. I certainly didn't know anything about Jung and had only a vague notion about psychology. But flying saucers sounded interesting.

The talk was a revelation, in that instead of speculations about little green men, Jung offered a sober description of the mentality of postwar Europe. The collapse of old systems of beliefs had created chaos in the minds of people—a churning anxiety, a yearning for some new order. It was this craving, he said, that led people to imagine they were seeing swirling disks in space—disks that represented the mandala, ancient symbol of spiritual harmony. Now, this may not have made a lot of sense, but it definitely touched on my own experience of the war and it made me sit up straight. I realized that there was a science I did not know about—a science that could perhaps explain why the adults I knew were so clueless. After returning to Italy, I read as many of Jung's books as I could find, then those of Freud, Adler, and a few of the other depth psychologists.

When I later came to the United States, mainly to study this strange new science, I was somewhat shocked to discover that

Jung and company were not considered very scientific after all, that whole armies of empirical psychologists had been busy, unbeknownst to me, plotting the laws of human behavior. My original infatuation with method revived, and I plunged with glee into the Cartesian delights of statistics.

At the University of Chicago, I befriended Friedrich von Hayek (he later earned a Nobel Prize in Economics in 1974), who used to go deer hunting with my grandfather before the war. He could not bear the mention of Jung in his presence. Instead one day he put a book in my hand and said, "Here, read this. If you want to be a scientist, this is all you need to know." The book, written by one of his friends, Karl Popper, was called *The Logic of Scientific Discovery*—or more literally translated from the original German, *The Logic of Research.*

In some ways that book became my bible, even though I could not fathom its second half, which was full of the intimidating formulae of formal logic. But the first hundred pages or so gave such a clear, convincing, and humane account of what scientific thinking entails that after reading it I never looked back again—or so I like to think, anyway—on the misty marshes where the fuzzy-minded live.

But perhaps the preference for empiricism and logic is in the genes. When my youngest son, Chris, was about seven years old, we had a friendly argument after dinner. He asked me if I had heard what he had been telling me about his day in school. After some attempts at covering up, I had to admit that I hadn't. He allowed that this was par for the course—that grown-ups rarely paid attention to what was going on. After chastising him for such an obviously unwarranted generalization, I felt the issue was settled.

A few days later, however, Chris asked me to look at a chart he had made. It had three columns, headed "Grown-ups," "Teens," and "Kids," and each column was further divided into two, labeled "Yes" and "No." There were long lines of check marks in each column. It turned out that each afternoon after school he had spent an hour or so at the living-room window, shining the bright reflection of my shaving mirror on the sidewalk three floors below. When a person approached a certain crack in the sidewalk, Chris would shine the beam about six feet in front of the pedestrian and keep moving it ahead until he or she reached another crack fifty or so feet down the sidewalk. If the person began to look for the source of the light, he would mark the "Yes" column in the appropriate age category; if the person seemed oblivious to it, he would check "No."

Having the tables thus turned on me, I had no choice but to show Chris how to perform a chi-square test, which showed beyond a shadow of a doubt that the large difference between "Kids" and "Grown-ups" in the proportion of "Yes" marks could have happened by chance only once in 10,000 tries; hence the reasonable conclusion was that kids were indeed more aware of what happened around them than adults were. Although Chris was vindicated (at least in part, since I of course did my best to talk myself out of the implications of the data), I also felt good knowing that the torch of empirical research had passed down to another generation.

If one were to generalize from my example and that of Chris, one might conclude that the impetus for social science is to prove someone wrong—one's friend, one's father. Perhaps so. But the beauty of it is that if rational rules are followed, some semblance of order begins to steer the discussion away from sheer force or rhetorical guile. As the years go by, I have come to realize the limitations of numbers, logic, and statistics; nevertheless, those early experiences inoculated me for good against the temptation to lapse into prejudice, superstition, or irrationality. I now feel toward the empirical method as Churchill felt about democracy: It may be a wretched system, but the alternatives are far worse.

My Father and Albert Einstein

MURRAY GELL-MANN

MURRAY GELL-MANN is a Distinguished Fellow of the Santa Fe Institute, the Robert Andrews Millikan Professor of Theoretical Physics Emeritus at the California Institute of Technology, a member of the National Academy of Sciences, and a Foreign Member of the Royal Society of London. In 1969 Professor Gell-Mann received the Nobel Prize in Physics for his work on the theory of elementary particles. He is the author of *The Quark and the Jaguar: Adventures in the Simple and the Complex.*

I was born on Manhattan Island shortly before the stock market crash of 1929, and I grew up there except for a few years in the depths of the Depression when my family's situation became especially difficult and we couldn't afford the rents in Manhattan. Not only did the crash herald the beginning of the Depression, but in addition the draconian National Origins Act of 1924 became fully effective in 1929. Both these developments were bad for my father, because he ran a small language school. A German-speaking immigrant from the Austrian part of Austria-Hungary, he had learned flawless English as a young adult. His pronunciation and grammar were perfect; you might suspect he was a foreigner only because he never made mistakes. Besides teaching English to immigrants, he taught German and hired other teachers for the Romance languages. His language school achieved a modest success. However, the combination of the Depression and the dearth of new immigrants did it in, and we fled from the Gramercy Park district to a neighborhood near the Bronx Zoo, returning to Manhattan, to the Upper West Side, around my eighth birthday.

Throughout these peregrinations, my mother had the idea that I was a little bit special, and she tried very hard to get me into a private school—although my father had no interest in that whatsoever. I didn't know what was happening, but I kept having to pile blocks on top of one another in various tests at different places in New York City. I realize now, of course, that these were attempts to get me into a private school with a full scholarship. They all failed, unfortunately, until finally a very nice music teacher named Florence Freint succeeded in getting me into Columbia Grammar School. We lived on West 93rd Street, almost across the street from the school. It had a long history, having been founded in 1764 as part of Kings College, which became Columbia University. I entered the sixth grade there in 1937, at the age of eight, on a full scholarship.

My brother Ben was a wonderful influence in my life. Ben was almost nine years old when I was born, and, like me, was three years ahead of most other students in his school. He taught me to read, from a cracker box, when I was three. He taught me almost everything I knew when I was little. Ben and I would do all sorts of things together. We played games and we visited museums. We loved bird-watching, and we were also interested in plants, butterflies, giant silk moths, and mammals. We still went up to the Bronx for some of our bird-watching after we moved back to Manhattan, because just north of the Bronx Zoo is the only remaining stretch of the hemlock forest that once covered the whole of New York. Ben and I regarded the city as a hemlock forest that had been overlogged.

At home, the atmosphere was always friendly to science. My father was very devoted to mathematics, physics, and astronomy. He tried to learn advanced physics, particularly general relativity, and was a great admirer of Albert Einstein. He never succeeded in fully understanding general relativity, but he had books on the subject and worked very hard at reading them. I didn't have an inordinate amount of interest in physical science, although I did like astronomy. I was more interested in natural history and in archaeology and linguistics. All those subjects involve complexity, diversity, and evolution, and they depend a good deal on historical accident as well as on fundamental principles.

I was a student at Columbia Grammar for seven years, from sixth through twelfth grade, and I decided that I would like to go on to Yale University, where one all-inclusive full scholarship was available. When it came time to apply for admission, I had to fill out a form that asked, among other things, what my major subject would be if I were admitted. At the time I thought it very unlikely that I would end up at Yale: First, I would have to get through its rigorous admission process; second, my parents couldn't contribute any funds at all, and so I would have to receive that one full scholarship. Nevertheless I was applying and found myself faced with this question. Somewhat uncharacteristically, I talked it over with my father.

"What were you thinking of putting down?" he asked.

I said, "Whatever would be appropriate for archaeology or linguistics, or both, because those are the things I'm most enthusiastic about. I'm also interested in natural history and exploration."

"You'll starve!" was his encouraging response. This was 1944, and his experiences with the Depression were fresh in his mind. We were still living in what might be described as genteel poverty.

"What would you suggest?" I asked.

He mentioned engineering, to which I replied, "I'd rather starve. Besides, if I designed anything it would fall apart." (Sure enough, when I took an aptitude test a year later I was told to take up anything but engineering.)

"Why don't we compromise—on physics?" he suggested. I pointed out to him that I had taken a course in "physics" at Columbia Grammar and that it was not only the dullest course I had ever taken but also the only one in which I had ever done badly. In it, we memorized the seven kinds of simple machines and then learned little bits about heat, light, electricity, magnetism, wave motion, mechanics, and so forth—with no hint whatsoever that these topics might be related to one another. I couldn't possibly major in such a subject, I protested.

"It will be very different when you take advanced courses in physics," my physics-besotted father insisted. "You'll learn general relativity and quantum mechanics. And those are very, very beautiful."

I thought I would please the old man, since it didn't really make any difference what I wrote down on this form. If, through some miracle, I was admitted to Yale and also got the one scholarship that would allow me to attend, I could always change my mind.

Well, things indeed turned out that way—but not quite. When I arrived in New Haven, I was too lazy to switch major subjects, so I actually took physics courses, among others, and got hooked on quantum mechanics and relativity, just as my father had predicted.

Incidentally, the scholarship that made it possible for me to attend Yale was called the Medill McCormick Scholarship. It provided for everything. I didn't even have to work at a "bursary job," like just about everyone else on scholarship. I was a little troubled about the names on the scholarship, though, because I knew that Bertie McCormick ran the *Chicago Tribune* and Joseph Medill Patterson ran the *New York Daily News*, neither of which was especially antifascist in the days leading up to the war. As a strong opponent of fascism—after all, the Second World War was still going on when I got to Yale—the idea that I was receiving money from a scholarship with those names on it was troubling to me.

Nothing happened, though, until 1947, when I went to see the scholarship authorities to ask if I could spend a ninth term at Yale, because the scholarship was about to expire. I was due to graduate in January 1948, but I preferred to stay until June, since I didn't think it would be possible to enter graduate school in the middle of the year. The scholarship people finally agreed to give me the ninth term but requested that I write a letter to my unknown benefactor. I struggled with the notion of writing this letter. At one point I composed a note that I had no real intention of sending, saying in essence, "I'm very grateful for your generosity in providing this money that allows poor boys to attend Yale and that has made an enormous difference to my future, but I'm somewhat troubled about the possible origin of the money." I went on to complain about the newspapers. Needless to say, and perhaps wisely, I never sent the letter.

Thirty years later, at a garden party in Aspen, I met a wonderful lady named Trini Barnes—Katrina McCormick Barnes. When she was young, her parents died and left her a good deal of McCormick wealth, but she had roughly the same attitude toward it as I did! She made her uncle Bertie pay through the nose for her shares of the Tribune Company, and then she started to give away her money. She was the "unknown benefactor," and the scholarship was named after her brother Medill, who passed away before he could attend a university. That letter would not have been as much of a disaster as one might have thought.

As it happened, the teaching of physics at Yale at that time was not particularly good, but there were a few exceptions. I was fortunate enough to have a course with Henry Margenau. He had completed his PhD at Yale in 1929, and although he didn't accomplish an enormous amount of research, he was a wonderful teacher. The course was called "The Philosophy of Physics," and he taught it on Tuesdays, Thursdays, and Saturdays at something like ten in the morning. The course didn't confine itself, however, to the philosophy of physics. It was really about physics itself, with some mention of philosophy in connection with each topic. I was happy to tolerate that much philosophy, and the teaching about physics was spectacular. I was only a sophomore and some of the students were juniors; we hadn't studied a great deal of theoretical physics. The advanced topics, like quantum mechanics and general relativity, loomed as difficult obstacles to overcome, but Margenau made everything miraculously easy.

We started with Lagrangian and Hamiltonian mechanics. Margenau began by saying, "I assume you've all studied the calculus of variations."

We said no.

He then asked, "What do your math teachers do in class today? Apparently they don't teach you anything useful, just epsilons and deltas [a reference to the highly rigorous approach to analysis often taken by professional mathematicians]. I'll teach you the calculus of variations today, and then on Thursday and Saturday we'll do Lagrangian mechanics." That's what he did, and it didn't take a moment longer than he said it would.

Then we went on to special relativity, and we got through that in a week or two. Next he announced, "We're going to work on general relativity now, but you need to know tensor analysis. I assume you all know tensor analysis?"

We said no.

"Well, that's a problem," said Margenau. "We'll have to do tensor analysis today and Thursday, and then on Saturday we'll start on general relativity."

And so it went, all the way through, including quantum mechanics.

In June 1948 I received my bachelor's degree in physics, and I planned to enter graduate school in physics in the fall. The results of my applications were disappointing. Harvard admitted me but offered no financial aid. Princeton turned me down flat. At Yale, I was admitted to graduate school in mathematics, but not in physics. The one encouraging reply from a physics department came from MIT. I was admitted and offered the job of assistant to a theoretical physics professor named Victor Weisskopf, of whom I had never heard. When I inquired about him, I was told he was a wonderful man and an excellent physicist and that everyone called him by his nickname, Viki. He wrote me a very nice letter saying he hoped I would come to MIT and work with him.

I was still discouraged, though, about having to go to MIT, which seemed so grubby compared with the Ivy League. I thought of killing myself (at the age of eighteen) but soon decided that I could always try MIT and then kill myself later if it was that bad, but that I couldn't commit suicide and try MIT afterward. The two operations didn't commute, as we say in math and physics jargon. When I got to MIT in the fall, I discovered that it was actually a very pleasant place, with agreeable fellow students (including many from the Ivy League), and a number of excellent professors, Viki, of course, among them. I had one of several desks in the large room next to Viki's office, and my officemates were changing all the time. They were not all graduate students—many already had PhDs, including Marvin L. "Murph" Goldberger, who had studied with Enrico Fermi at the University of Chicago. Murph and I became friends and talked a great deal about physics and other subjects. He left to go back to Chicago as an assistant professor. Later he would get me a job as an instructor there.

I should have gotten my PhD at MIT in a year and a half, but I unfortunately dragged out the writing of my dissertation. I spent a lot of time reading things like Walter Yeeling Evans-Wentz's translation of the *Tibetan Book of the Dead.* I finished in January 1951, about seven months late. I was supposed to start a postdoctoral year at the Institute for Advanced Study in September 1950, but because of the delay my year at the institute began in January 1951. I was now twenty-one, and I lived in a rooming house just across the street from Princeton University.

Albert Einstein, my father's hero, was also at the institute, and he came to work regularly. I could have spoken with him, and that certainly would have impressed my father, but at that time I didn't like the kind of people who approached great figures, introduced themselves, got into conversation with them, and reported the experience to others as a way of showing off. So I didn't approach him.

It is interesting, by the way, that people everywhere have chosen Einstein as a symbol of greatness in physical science. It needn't have turned out that way; they could perfectly well have fixed on someone much less distinguished. But Einstein really was a great genius in theoretical physics and fully deserved the adulation he received from the public. At the time, he was working on his attempt to construct a unified field theory. The general idea of seeking such a theory was of course an excellent one, but the way he was going about it clearly doomed the work to failure. He didn't believe in quantum mechanics, so his theory was purely classical. Nor did he introduce elementary particles like the electron, hoping that they would somehow emerge from his equations. Also, he included only the electromagnetic and gravitational fields, omitting the other known fundamental forces of nature, the strong and weak interactions.

If he had been working on something that looked promising, it would have given me a perfectly legitimate reason to talk with him and I would probably have done so. However, as it was, I said "Good afternoon" occasionally but nothing more. Einstein would answer in what seemed like a mixture of German and English, and that was it. Today I would almost certainly behave differently, asking the old man about his thought patterns of years before, when he was carrying out the greatest physics research since Newton's. That would have been exciting! But asking him about his life and his attitudes toward the world and toward physics was not something I felt comfortable doing at the time. Nowadays, older and somewhat wiser, I would probably not let such an opportunity slip by.

A Midcentury Modern Education

ALISON GOPNIK

ALISON GOPNIK is a professor of cognitive science in the psychology department of the University of California at Berkeley. She is the coauthor of *Words, Thoughts, and Theories* and *The Scientist in the Crib: What Early Learning Tells Us About the Mind.*

Henry James once said that the James family was his native country and he knew no other. This is true of many close, large, talkative families, like the Jameses, and it was certainly true of the close, large, talkative Gopnik family. My parents, my five younger brothers and sisters, and I lived in our own native country, encircled though we were by the barbarous nation of 1960s Jewish middle-class Philadelphia. Our founding fathers were Galileo, Darwin, and Einstein; Shakespeare, T. S. Eliot and James Joyce; Picasso, Marcel Duchamp, and Mies van der Rohe. Science, Art, and Modernity were our constitutional principles, and Superstition, Vulgarity, and Sentimentality were our deadly enemies.

In retrospect, I realize that my parents were part of a much broader postwar social revolution. The traditions of high European modernism, the sensibilities of the Vienna Circle, Montmartre, and Weimar, were transplanted after the war to the children of poor Jewish immigrants in Philadelphia, Newark, and Brooklyn. My grandfather ran a corner grocery store in a rundown Philadelphia neighborhood and never finished elementary school. But my parents, like an entire generation of American intellectuals, discovered the booming postwar public libraries and museums and concert halls and ended up populating the booming American universities, first as scholarship students and later as professors. The comedy of Woody Allen and Philip Roth, who are my parents' contemporaries, is the comedy of this transformation; bagels and knishes meet Flaubert and Kafka. My parents had the special enthusiasm of a generation who felt that they had created a new world all by themselves.

But my experience as the child of these children of immigrants was a bit different. All children see their parents *sub specie aeternitatis*—as unchanging, eternal elements of the natural world, rather than as particular people at a particular phase of history during a few particular years in their lives. And this is especially true in big families where all the background is shared. So for us kids the passionate intellectual life was just the way things were.

Historical context or no historical context, the intellectual enthusiasm in our house was idiosyncratically intense. In 1959, when I was four and my brother was three, my mother dressed us in gold velveteen outfits sewn for the occasion, and we drove to New York in our battered VW bug and stood in line for hours for the opening of the Guggenheim Museum. We walked through the museum, got in the car, and drove back to Philadelphia. (My parents disapproved of the architecture; they thought Frank Lloyd Wright was a little sentimental.) At around the same age, we went trick-or-treating as Hamlet and Ophelia, in handmade wool wigs. For later Halloweens, my siblings and I dressed up as the Greek pantheon (I was Athena), an abstract-expressionist art gallery (I was a Franz Kline), and a fully articulated four-child-long version of the dragon in *Beowulf* (I was in the head and orchestrated both the flames and the smaller children behind me). When I was ten, four of us appeared every night in Bertolt Brecht's *Galileo,* directed by Andre Gregory, then a young avant-garde director starting out in the Philadelphia boondocks.

Other families took their kids to the theater to see *The Sound of Music* or *Carousel;* we saw Racine's *Phaedra* and Samuel Beckett's *Endgame.* (My parents thought *Waiting for Godot* was a little sentimental.) Other families went to see the Statue of Liberty and the Empire State Building when they visited New York; we went to see Lever House and the Seagram Building. Other families listened to Beethoven or the Texaco–Metropolitan Opera radio broadcasts; we listened to Alfred Deller singing John Dowland, and Robert Craft's recordings of Gesualdo motets. Other little girls had page-boy haircuts and wore white socks and black patent-leather Mary Janes. My sisters and I all had very long, straight, dark hair, and my favorite outfit was a black leotard and tights under an olive-green jumper made of tent canvas. Our family read Henry Fielding's eighteenth-century novel *Joseph Andrews* out loud to one another around the fire on camping trips, and when we decided, like all those other kids, to put on a play, we chose the great screen scene from Sheridan's *The School for Scandal.*

The artistic and literary sides of our lives were most flamboyant; we were a theatrical family in more ways than one. But it was a mark of my parents' brand of modernism that science played an integral role in their vision of high culture. We were proud—and, indeed, practically religious—atheists, and the story of Galileo's persecution by the Inquisition was one of our founding mythologies. "And still it moves" was one of our catchphrases. My father became an English professor, but he took courses at Penn with Nelson Goodman, the great logical empiricist philosopher, and told us about the principles of verificationism and the problem of induction. My mother was in one of the first serious linguistics programs, also at Penn, just as Noam Chomsky, a fellow student a few years earlier, was revolutionizing the study of language. "The program is in formal linguistics and mathematical logic," she would say coolly when some rug-selling cousin asked how many languages she spoke. When my fourth-grade teacher defined a noun as a "person, place, or thing," I had to explain to her about distributional analysis and transformational grammar. (We alternately ignored and despised school, but at this remove one can't help feeling a bit sorry for the hapless teachers who found themselves with omniscient Gopnik children in their classes.)

We were weird, no question about it. We were "precocious children," "child prodigies," and, according to the pop psychology of the day, we should have been twisted neurotics. But the truly extraordinary, really weird thing about our upbringing was my parents' gift for making this weirdness seem absolutely natural and normal—the accepted, ordinary, happy way that civilized people went about their lives. They were devoted to their children's intellectual lives all right, but their devotion was utterly unlike the twenty-first-century, upwardly mobile, middle-class parental obsession with "enrichment" and "achievement."

My parents met when they were eighteen, dropped out of college, and had six children in the next eleven years. When I was a child, my father worked several clerical jobs, often at once, and my mother took care of us, while they both went back to school and got PhDs. We never had much money, and when I was growing up we were positively poor, though my parents somehow always managed to surround us with good art and well-designed modern furniture. Until I was five, we lived in a public housing project. Then my parents bought a big dilapidated old Victorian house at 41st and Locust Streets (then a very dubious neighborhood) for $9,000 and knocked down walls, exposed brick, and painted everything white, long before that was fashionable.

We went to ordinary Philadelphia public schools and never were in a "gifted" program or an after-school class or summer camp. In school, during the hopelessly dull classes, I hid a book under my desk and read. After school, I came home and curled up in the midcentury-modern Bertoia chair and read. In the summer, I sat in the garden in the Eames chair and read. After elementary school I never got particularly good grades, and this was never a big deal. I would not have been admitted to Berkeley, where I now teach. For me, intellectual life wasn't something you achieved, it was something you breathed. I never felt "enriched," though I did sometimes notice that other kids seemed oddly, peculiarly, impoverished. And I was very happy.

I never decided to become a scientist. I did know that I wanted to be a philosopher and spend my life thinking, and that I wanted to think about children. Becoming a developmental psychologist just turned out to be the best way to do that. Philosophy, particularly rigorous analytical philosophy of the Nelson Goodman sort, was very much part of the background of our house. But then just about every discipline was part of the background of our house. My little brothers and sisters grew up to be a *New Yorker* writer, the head of the National Academy of Sciences' ocean studies board, a Near Eastern archaeologist, the *Washington Post* art critic, and a public-health manager. I feel sure that each of them could find the roots of their vocation somewhere in the polymath atmosphere of the Gopnik household.

I was an omnivorous reader and devoured science books, but then I devoured everything. I read the wonderful George Gamow Mr. Tompkins books, with their vivid visualizations of quantum theory and relativity, which still provide me with my mental images of physics—I still think of electrons as naked, mustachioed, middle-aged men whirling through space. I read Eve Curie's biography of her mother, *Madame Curie*, many times over. (I wonder whether there is any woman in science today who didn't read *Madame Curie* many times over.) But I wasn't the sort of kid who did chemistry projects or collected beetles, and while I liked to read about Marie Curie painstakingly distilling the giant heap of pitchblende to find the teaspoonful of radium, I thought it sounded like rather tedious work. (It was only in graduate school, when I switched from philosophy to psychology, that I discovered that distilling the pitchblende to find the radium—in psychology, we use control conditions and pilot studies—could actually be fun.) My favorite science book was *Discovering Scientific Method*, a book that was not about any science in particular but explained the philosophy of science and showed how to apply scientific thinking to everyday life.

My first memorable encounter with philosophy, though, came—oddly enough—through TV. One night when I was ten we watched a dramatization of Socrates' last days, and though my parents thought it was a little sentimental, I loved it. I have always wondered how Socrates got on prime-time TV in 1966, and as I was writing this essay I e-mailed my family to see if they remembered any more details. My writer brother Adam immediately reported that what we had seen was actually a play called *Barefoot in Athens* by Maxwell Anderson, with Peter Ustinov in the lead. He remembered the drama, but I remembered only the argument and had entirely erased the literary details (though this does explain why I have always imagined Socrates with a blond beard and an English accent). It wasn't so much the heroic story I liked as the idea that these people did nothing but think and talk all day.

Characteristically, my parents handed me a copy of the Penguin collection of Plato's works, a battered paperback with Raphael's famous picture of the Academy on the cover; they never thought that anything was too difficult or grown-up for their children to read. This, I decided, was how I wanted to live. Ancient Athens, like eighteenth-century London and 1920s Paris, was clearly a province of Gopnik country, and I made models of the Acropolis in the backyard with sticks and rocks figuring as arguing philosophers.

But even in that very first encounter with philosophy there was a catch. The argument in the Penguin Plato that impressed me the most was Socrates' case, in the *Phaedo,* for immortality and against death. Like practically all ten-year-olds, precocious or not, I worried a lot about death, and as a committed atheist I was certainly in the market for a good argument for immortality. Socrates argues that something as complex as the soul can't appear and vanish out of nowhere, and therefore it must exist, before and after our individual lives, in an abstract Platonic heaven. What struck me about the argument was that there was no mention anywhere of children. It seemed obvious to me that your soul was created at least in part by the genes you inherited and the ideas you acquired from your parents, and that it continued after death in the genes and ideas you passed on to your children. Of course, this idea depended on scientific concepts that weren't available to

Socrates. But what really struck me was that even if Socrates didn't know about genes, he must have known about children, and yet they were never even mentioned in the *Phaedo.*

This seemed particularly strange because—in our family, and for me in particular—children were so obviously important and interesting. Like many oldest sisters, I was an unofficial parent (according to my younger siblings, considerably stricter and bossier than the official ones). I had my own first child soon after I left home, and my youngest is still only fifteen. I have never lived a life in which I was not taking care of children, and babies and young children have always seemed to me to be the most surprising, unpredictable, and interesting of companions.

Many scientists report that they first became fascinated by their subject matter long before they understood much about science as an institution—all those child stargazers and butterfly collectors and dinosaur maniacs. I suspect that there are few reports of scientists with a childhood fascination for babies, because most of those children turned into nursery school teachers or children's librarians or just stay-at-home mothers. An intelligent girl who was fascinated by stars might well brave the hurdles that face a woman in science. What else could she do? An intelligent girl who was fascinated by children was behaving just the way girls were supposed to behave and wouldn't even have thought about science as a possible career. And this is particularly true because children are actually the biggest hurdle facing women in science. Scientific institutions make it very difficult for women to combine child rearing and a scientific career. It seems to me now that I was destined to become either a psychologically minded philosopher or a philosophically minded psychologist. But given slightly different contingencies, I might have become a frustrated preschool teacher or faculty wife.

Just when childhood ends is a difficult question, especially for precocious children who (depending on your point of view) have always been grown-up or never grow up at all. For me, there was a sharp dividing line at twelve, when we moved from Philadelphia to Montreal and I entered a mercifully short period of high school misery. But I was still only fifteen in 1970 when, through a weird combination of Quebec politics and bureaucratic fluke, I entered McGill University. My parents had faculty jobs there by then, and I learned early that I could take any course I wanted to, if I was determined enough. My freshman year was an annus mirabilis. I attended a graduate seminar that combined faculty from philosophy, computer science, psychology, and linguistics—it must have been one of the first cognitive science courses anywhere in North America—and became a philosophy honors student and founder, president, chief activist, and head cook of the Philosophy Students Association. McGill might not seem an obvious home of the cognitive revolution, but there was certainly something in the air; a remarkably large number of my undergraduate contemporaries there went on to become accomplished cognitive scientists.

I also had a lot of other kinds of informal tuition in my teens. I suppose I agree with policies that ban undergraduates from sleeping with their teaching assistants, but "Introductory Syntax," "Philosophical Logic," and "The Psychology of Language" would have been a lot duller if those policies had been in force at McGill. For about five minutes in the seventies, after the Pill but before AIDS, it was possible for fifteen-year-old girls to experience genuinely carefree sexual adventure for the first and only time in history. I'm glad I was fifteen then. I also agree in general with policies that ban minors from bars, but I'm glad that they weren't in force at the legendary Le Bistro on Mountain Street. Students like me—and artists and journalists, scientists and political activists—drank Pernod and Labatt's 50 beer at the zinc bar, smoked evil-smelling Gauloises and Gitanes at the marble tables, and talked late into the cold Montreal winter night. There was a general glow and excitement and sense of possibility in Montreal then, in the bars and in the streets as well as in the classrooms. Politics and philosophy, sex and science, all seemed expressions of the same underlying revolutionary spirit.

As a developmental psychologist, I am often called upon to pontificate about how science education in America could be reformed. There were aspects of my own education that could perhaps have been improved. I might have been a better scientist if I had learned to be more disciplined and hardworking—less of a fox and more of a hedgehog. I certainly wish I had been forced to do more mathematics. But in general I think the education that most children get should be more like my informal education. My siblings and I weren't prodigies by nature. We were ordinary children who had rich opportunities to learn and who were taken seriously by people who cared about us. I think preschools and elementary schools should be much more like that big Victorian house with the modern furniture on Locust Street, and high schools and universities should be much more like Le Bistro, though perhaps with somewhat less smoke and alcohol. And I think young women shouldn't feel that they are defying the odds if they try to combine motherhood and a scientific career. I was lucky, but children—and science—shouldn't have to rely on luck.

Cosmology Calls

PAUL C. W. DAVIES

PAUL C. W. DAVIES is a professor of natural philosophy in the Australian Centre for Astrobiology at Macquarie University, Sydney. His books include *The Fifth Miracle, About Time, The Mind of God,* and *The Last Three Minutes.*

I was born to be a theoretical physicist. I know it sounds old-fashioned, but there is such a thing as a calling. And I had it—I still have it. From the earliest age I can remember, I wanted to follow the path I later chose when I eventually understood what theoretical physics was and how professional science works. There was no epiphany. No key event precipitated my decision; no mentor inspired me.

My family thought I was nuts. Science didn't figure anywhere in the Davies genetic heritage. "When are you going to get a proper job?" an aunt asked me at a family wedding just after I had secured my first lecturing position, at King's College in London. "What, exactly, is physics?" my grandmother wanted to know. My father, ever practical, was skeptical about a life spent contemplating the mysterious workings of the cosmos. "Nobody will pay you just to sit and think," he declared. My mother would have approved of my becoming a scientist had I worked on a cure for cancer, but physics seemed mysterious and vaguely sinister to her.

So how did a normal, fun-loving suburban London kid end up as a theoretical physicist and cosmologist?

There has always been something deep inside me—a sort of restlessness, verging on a sense of destiny—that drives me. It's a feeling of being drawn inexorably toward the serene heart of existence, a compulsion to search for hidden meaning in the universe, along with a conviction that meaning is in fact out there, lying just—but only just—within my grasp.

Of course I am not alone in experiencing such intimations of cosmic significance, but most people outgrow them. Others relieve themselves of the burden by turning to mysticism or religion. I flirted with conventional religion in my midteens but found it disappointingly shallow, the answers either too glib or else incomprehensible. I remember at the age of sixteen being gravely troubled by the paradox of free will. Why didn't the atoms of my brain just do what atoms have to do, irrespective of what I *wanted* to do? And where did my wants come from anyway? Even if I could somehow magically do what I wanted, how could I *want* what I wanted?

I discussed these concerns with the curate at the local Anglican youth club. I had joined this organization primarily in search of a girlfriend and a better social life, but I did enjoy the occasional deep and meaningful chat with the clergy. These conversations tended to focus on sex and why it wasn't a good idea, at least among the youth club membership. Free will wasn't high on the agenda, and my anxieties were shrugged aside. But the encounter was a turning point for me. It struck me then that the best way to answer not only the puzzle of free will but all the Really Big Questions of existence was not through religion but theoretical physics. Physics is, after all, the tune to which the atoms of my brain dance.

So theoretical physics became, in a sense, my religious quest, the best hope I had of making sense of the world and my place within it.

But why did this matter so much to me? Why didn't I just ignore this existential crisis and get a "proper job," like my peers?

Part of the answer was boredom, sheer mind-numbing boredom. I grew up in postwar north London. Toys were scarce, food monotonous, school dreary. Trips to the seaside were a rarity, nobody owned a television, and even books were a luxury. Nothing exciting ever seemed to happen. I longed for some truly dramatic event—an alien invasion, a ghostly visitation, a message from the dead—anything that would relieve the daily tedium. Science became my escape. In the winter of 1955, when I was eight years old, my father took me as a special treat to see *The Benny Goodman Story* at the local cinema. We walked back home in the dark, through a small wood, and my father pointed out the bright star Sirius and some well-known constellations. I remember vividly the sharp points of light in the blackness of the sky, seen through the skeletal, leafless trees. Then we saw a shooting star. I had already noticed these fleeting objects from our back garden but had taken them to be a strange form of fireworks. My father explained that they were meteorites plunging into Earth's atmosphere.

Now this was pure magic! Merely by looking up, I could escape into a wonderland of literally otherworldly objects. It was all there just above my head, yet it went unnoticed by most of the people around me, preoccupied as they were with their ordinary affairs.

From then on, I was hooked on science. Light and electricity thrilled me especially. I still remember how weird it felt to see the image of my bedroom window projected through a pinhole I punched in a shoe box. Or the sense of achievement in getting flashlight bulbs to glow mysteriously from complicated combinations of wires and batteries. What most captivated me about these rough-and-ready experiments was the discovery that a hidden realm of forces and phenomena could be accessed with a heap of junk in my own bedroom, as long as I figured out the right procedures. So I sequestered all the bric-a-brac I could find—old scratched lenses, metal tubes, discarded firecrackers, frayed electrical wire, lightbulbs purloined from buses on the way home from school. And always with the thought "What can I make this do?"

For my twelfth birthday, my parents bought me a photographic developing kit. I didn't actually have a camera, but I borrowed my father's Box Brownie and tramped out in the snow at night to photograph streetlights and moving cars. I can still recall the eerie sensation when the image of our front door, illuminated by an overhead light reflected from the snow, emerged on the printing paper in the tray of developer. As I gently rocked the tray, I felt like a magician manipulating dark forces.

By age fourteen, I was determined to build a telescope. Buying one was unthinkable in those frugal times. I did buy a four-inch mirror, but I put the rest together myself from odds and ends. The tube was made of rolled-up linoleum, the mount from scavenged bits of wood and some nuts and bolts I procured from family and friends. It was enough to get me started at observational astronomy. Encouraged by this success, I decided to build a bigger instrument, which meant grinding my own eight-inch mirror. To this end I commandeered the kitchen and set about making a pitch lap, to be covered in carborundum powder. After many hours of arduous grinding on the lap, followed by a prolonged polishing operation with optical rouge, the mirror was ready for testing and its final figuring. I carried out the optical tests in the living room, with flashlights, razor blades, and other impromptu devices. Eventually, after much effort, I had a serviceable telescope mirror, mounted in a wooden tube nearly eight feet long, with a cradle to house it in. I took over a corner of the garden and erected the giant tube, which resembled an artillery piece, on a great chunk of concrete that I had laboriously mixed and cast, embedding metal bolts in it at carefully measured locations. The final instrument worked, but not as well as it might have: It needed a machined metal mount, not the wooden lash-up that was the best I could muster. Nevertheless, it was good enough for observing the moon and planets, if I took sufficient care not to rock the tube while peering through the eyepiece. I still have the mirror in my garage.

The culmination of my career as a teenage astronomer came in 1963, when Margaret Thatcher gave me a copy of Norton's Star Atlas. She was the member of Parliament for Finchley, where I went to school, and this was an end-of-year prize for good science results. Many years later, when I received the Templeton Prize (awarded for my work linking science and religion), Baroness Thatcher was one of the judges and kindly agreed to sign the atlas for me once again.

My early fascination with science wasn't restricted to astronomy. I was always intrigued by how things move. By the age of sixteen I had read Einstein's little book on relativity and knew about Mach's principle and other mysteries of space, time, and motion. I tried the patience of my long-suffering physics master with a succession of designs for perpetual motion machines, but this interest in dynamics was by no means purely abstract. For years I had loved making bows and arrows, spears and catapults. I fooled around with cannons made from cigar tubes packed with gunpowder from dismembered firecrackers, marbles serving as bullets. They never worked. On a trip to Paris, full of anticipation, I launched a paper plane from the top of the Eiffel Tower, whereupon it was instantly snatched by the wind and disappeared upward into the overcast.

A visit to the circus at the age of ten left a deep impression on me. One of the acts consisted of a blindfolded knife-thrower hurling flaming blades at a spread-eagled and scantily clad lady assistant, who was not only strapped to a circular board but also set in rotation. If I couldn't make it as a physicist, then a job executed with that sort of panache would suit me fine. I duly practiced in the backyard with my brother's scout knife, using the door of our potting shed as a target. I had a lot of trouble getting the knife to arrive at its destination blade first. (I still can't do this reliably.) Cowboys in the movies always seemed to be throwing knives across crowded saloons, with the tip obligingly burying itself in the overturned table with a satisfying thud. Mine kept hitting the door side-on and dropping to the ground.

When I finally (almost) got the hang of knife throwing, I enlisted a girl named Elizabeth, who lived across the street, as my female assistant. This game young lady stood with her back against the shed door while I took aim. (I had decided against the blindfold.) Elizabeth is still alive and well, and has fond memories of her early role in show business; in fact, she went on to become a famous dancer and stage actress, starring in the circus musical *Barnum*. So there we are: Maybe I can take some of the credit.

By the age of eighteen, the fun and games were over and I was engaged in the serious business of converting my childhood dream into professional reality, which meant enrolling for a university course. All physicists must make a crucial decision of whether to follow theory or experiment. From what I have said about telescope making and knife throwing, you might have imagined I would choose experimental physics. But I found lab work insufferably dull. Boredom set in very quickly. Things progressed too slowly, and mostly my experiments, like my paper-plane flying and my forays into artillery, didn't work. I lacked the patience or skill to design the apparatus properly and collect the data with enough accuracy to mean anything. I told my fellow students I suffered from laboraphobia.

My career in experimental physics came to an inglorious end during a second-year university practical examination. Students were let loose in the laboratory for an entire day and given the task of measuring something. In my case, it was the viscosity of water (*dull, dull!*). We were given no instructions. The challenge was to devise your own experiment with the equipment provided. This consisted of a steel cylinder, a flask, a stopwatch, a metal stand, a length of cotton, a little mirror, some plasticine, an electric light that projected a beam, and a translucent horizontal strip with a scale marked on it. After a while, I twigged to what was wanted. I suspended the cylinder from the stand with the length of cotton and immersed it in the flask, which was filled with water. The little mirror I fixed to the cotton with the plasticine and reflected the beam of light from it onto the horizontal strip. I then set the cylinder into a twisting motion ("torsional oscillation," to use the technical jargon), and followed its progress by monitoring the spot of light as it swung back and forth on the scale. The drag of the water had the effect of gradually damping the oscillations, so that by timing the decay of the swings, you could work out the viscosity of the water. Simple. Except in my case the cotton broke and the metal cylinder plunged to the bottom of the flask, shattering the glass and flooding the bench. Whatever inconclusive data I had by then recorded was obliterated. I chose the theoretical option thereafter.

In any event, theory better suited my temperament and conformed to my long-standing quest for meaning. I had discovered the charm of pure theory some years earlier, while still at Finchley. I had taken a fancy to a dark-haired girl called Lindsay, who was studying only the humanities and so spent long hours in the school library reading English literature. I contrived to sit opposite her one day, charged with the homework task of computing the trajectory of a ball projected up an inclined plane. As I was partway through several sheets of mathematics, the ravishing Lindsay looked across at me with a mixture of admiration and puzzlement. "What are you doing?" she asked. When I explained, she seemed completely mystified. "But how can you tell where a ball will go by writing squiggles on paper?"

Lindsay's question has haunted me ever since. How is it, indeed, that we can capture the workings of nature using human mathematics? I came to see the equations of theoretical physics as the universe's hidden subtext. By learning the arcane language and procedures of mathematics, I could access an occult world of forces and fields, of invisible subatomic particles and subtle interactions—a wonderland at my fingertips every bit as compelling as the dark night sky above our heads but somehow more exciting because of its abstract nature. I felt as if I had been inducted into a secret society, where by following a set of special rules I could unveil an alternative reality—in fact, a deeper level of reality, which somehow came closer to the soul. The soul of the universe, perhaps. I realized then how Galileo must have felt when he wrote that the book of nature is written in mathematical language, and I experienced something of the same thrill: the sense that nature itself was speaking to me, in code.

People are fond of telling nostalgic tales about how they wanted to be a train driver or a brain surgeon or an astronaut when they were in their teens. I always wanted to be a theoretical physicist with a slant toward cosmology, and that's what I did. Looking back, I can't imagine what else I would have done. The hardest part was enduring the other obligatory subjects on the way. What was the point of learning chemistry or English? How would they help? I wanted to get right on with the main game. To that end, I was assisted by the narrow British curriculum that fast-tracked single-minded budding scientists to their goal. By eighteen I was studying only physics and mathematics; by twenty I was specializing in theoretical physics. I completed my PhD thesis before my twenty-fourth birthday and I was ready to take on the universe. Like most scientists, I still look at the world with wonder and ask myself, "What's it all about?" One day I'll know.

Member of the Club

FREEMAN J. DYSON

FREEMAN J. DYSON is professor emeritus of physics at the Institute for Advanced Study, a member of the National Academy of Sciences, and a Fellow of the Royal Society. He is the author of a number of books about science for the general public, most recently *Imagined Worlds* and *The Sun, the Genome, and the Internet.*

I was not driven to become a scientist by any craving to understand the mysteries of nature. I never sat and thought deep thoughts. I never had any ambition to discover new elements or cure diseases. My strong suit was always mathematics; I just enjoyed calculating, and I fell in love with numbers. Science was exciting because it was full of things I could calculate.

One episode I remember vividly. I don't know how old I was; I only know that I was young enough to be put down for an afternoon nap in my crib. The crib had solid mahogany sidepieces so that I couldn't climb out. I didn't feel like sleeping, so I spent the time calculating. I added one plus a half plus a quarter plus an eighth plus a sixteenth and so on, and I discovered that if you go on adding like this forever you end up with two. Then I tried adding one plus a third plus a ninth and so on, and discovered that if you go on adding like this forever you end up with one and a half. Then I tried one plus a quarter and so on, and ended up with one and a third. So I had discovered infinite series. I don't remember talking about this to anybody at the time. It was just a game I enjoyed playing.

Another memorable episode is the total eclipse of the sun in the summer of 1927. I was then three and a half. My father said that the eclipse would be total at Giggleswick, in Yorkshire. We lived in Winchester, about two hundred miles south of Giggleswick. My sister and I saw the partial eclipse at Winchester, looking at it through bits of glass that had been made dark by holding them over a smoky candle flame. We saw the sun slowly shrinking until it became a narrow crescent. But I was furiously angry at my father because he refused to take us to Giggleswick. I asked him when the next total eclipse visible in England would be; "1999," he replied. I calculated that I would have to live to age seventy-five to see the next eclipse. That made me even angrier.

Some years ago, after my mother's death, I found among her papers some relics she had preserved of my childhood. One of them is a paper headed "ASTRONIMY," with a sentence for each of the planets. For example: "You can hadly ever see Murcery becose the Sun is nearly allways in frount of it." At the bottom of the page, in my mother's hand, is: "FJD aged five and a half." This scrap of paper is evidence of two things: First, I had a mother who cared and encouraged me to learn; second, the fact that the statement about Mercury is wrong shows that I did not copy it from grown-ups. I must have made it up.

My father was a musician and my mother was a lawyer, neither of them expert in science. But they read the current popular science books by A. N. Whitehead, Arthur Eddington, James Jeans, Lancelot Hogben, and J. B. S. Haldane, and left them on the shelves for me to browse. Among them was Eddington's *Space, Time, and Gravitation*, an excellent introduction to relativity. I still have my father's copy, with his signature showing that he bought it when it was published in 1920. On page 49 there is a diagram of spacetime, with space plotted horizontally and time vertically, the light cones appearing as two perpendicular lines running diagonally and dividing spacetime into four regions. The text explains that the region above the diagonals is the absolute future and the region below the diagonals is the absolute past, while the two regions at the sides, between the diagonals, are "elsewhere." When I was seven years old, *Punch* published a cartoon of a nanny talking to a small boy who is lying on the ground with a book. The boy is me. The book is Eddington's *Space, Time, and Gravitation.* The nanny says, "Do you know where your sister is?" and the boy answers, "Somewhere in the absolute elsewhere." I remember being puzzled when my father said he was sending that conversation to *Punch* to be published. I didn't see the joke. What I had said was true, not funny.

In that same year, 1931, the asteroid Eros came unusually close to the earth. This is the asteroid that the NEAR (Near Earth Asteroid Rendezvous) spacecraft explored and landed on in 2001. It is the biggest of the asteroids that regularly come close to us. In 1931 there was much talk of the terrible damage that Eros could do if it should collide with our planet. I listened to my parents talking about this at the breakfast table. My father told me that the Astronomer Royal, Sir Frank Dyson, was leading an international effort to observe Eros and calculate its orbit precisely. This was important, he said, because it would give us a more accurate measure of the distance between the earth and the sun. Sir Frank was not related to us, but he came from the same part of Yorkshire as my father, and my father knew him. I liked the idea of calculating orbits precisely, and I thought that maybe I would one day be Astronomer Royal and calculate orbits too. Among the papers my mother preserved is a fragmentary story I wrote at the age of nine, entitled "Sir Phillip Roberts's Erolunar Collision," about an astronomer who calculates the orbit of Eros and discovers that it is heading for a collision with the moon. He predicts that the collision will happen in ten years, enough time for him to organize an expedition to the moon to observe the collision up close. At that point the story stops. Reading it again after seventy years, I find it interesting that Sir Phillip makes his great discovery by calculating and not by observing.

At the age of eight, I was, like other middle-class English kids at that time, sent away to boarding school. The school was a Dickensian horror, but it had one redeeming feature: a library where I could escape from sadistic boys and a sadistic headmaster. In the library was the Book of Knowledge, a popular children's encyclopedia, and the science fiction novels of Jules Verne. I read "Hector Servadac" and other Verne stories. I found the Verne puzzling at first, because I didn't know it was fiction. I thought Hector Servadac had really taken a ride on a comet and visited a new planet he called Gallia in honor of his native country. I was puzzled because Gallia did not appear among the planets in the Book of Knowledge. When I found out that the Verne stories were fiction, it was a big disappointment; I liked the Book of Knowledge better, because I could trust it. I read about the Yosemite Valley and pronounced it to rhyme with "nose bite." I read about matter being made up of electrons and protons. Then I read a long piece about electrons and electricity and electric motors, but there was nothing comparable about protons. I wondered why this was: Why didn't we have proticity and protic motors? I asked some of the boys and some of the teachers, but nobody knew. The school taught mostly mathematics and Latin. No science was taught. That was probably a good thing, as it made science more attractive to misfits like me. With a small group of friends I started a science society, circulating books and holding occasional meetings to discuss them.

At age twelve, I moved to Winchester College, a high-class private school where my father was the head music teacher. There I had the enormous luck to find three kindred spirits of my own age, James Lighthill and the brothers Christopher and Michael Longuet-Higgins. All four of us later became fellows of the Royal Society: James had a brilliant career in fluid dynamics, Christopher in theoretical chemistry, Michael in oceanography. James and I were both in love with mathematics and rapidly devoured the books in the school library. We read, for example, *Men of*

Mathematics, by Eric Temple Bell, a collection of romanticized biographies of great mathematicians. Bell was a professor of mathematics at the California Institute of Technology and a gifted writer who wrote with authority about mathematics and knew how to enthrall susceptible teenagers. His book turned a whole generation of young people into mathematics addicts. Although many of its details are historically inaccurate (for example, the chapter "Genius and Poverty" portrays the Norwegian mathematician Niels Henrik Abel as a starving genius, but a recent study of his expenses while living in Paris shows that he spent considerable sums on theater tickets), the important things are true. He portrays mathematicians as real people, with real faults and weaknesses ("The lives that follow will suggest that a mathematician can be as human as anybody else—sometimes distressingly more so"), and mathematics as a magic kingdom that people of many different kinds can share. The message for the young reader is, If they could do it, why not you?

Together we worked through the three volumes of Camille Jordan's *Cours d'analyse*, the famous textbook of nineteenth-century mathematics as it was taught to students at the elite École Polytechnique in Paris a hundred years ago. We could not have had a better introduction to serious mathematics. It went far beyond anything our Winchester teachers knew or cared about. We did not ask how it happened that this treasure was in our school library. Later we suspected that it must have been put there by the famous mathematician G. H. Hardy, who had been a boy in the same school forty years earlier.

The teachers at Winchester wisely left us alone to educate ourselves. They gave us a lot of freedom, confident that we would use it responsibly. In our last year at Winchester, we spent only seven hours a week in class. That summer I met my first real mathematician, a young man named Daniel Pedoe, who was a junior lecturer at Southampton University and whom the Winchester school authorities had hired to come once a week and give me private lessons. The sessions with Pedoe were a revelation. He had been a research student in Rome and a member at the Institute for Advanced Study in Princeton. He knew personally the legendary figures of mathematics, and he knew the latest fashionable stuff they had been doing. His own field was geometry. He gave me a German translation of Francesco Severi's *Algebraic Geometry* to study. Pedoe and I became lifelong friends, and Severi's book became one of my prized possessions. I did not become a geometer, but I acquired from Pedoe a taste for the geometric style that makes mathematics an art rather than a science. Pedoe later became a professor at the University of Minnesota and a leading spokesman for geometry in the international mathematical community. With his friend Hidetoshi Fukagawa, he published a book, *Japanese Temple Geometry Problems,* about an elegant indigenous version of geometry that flourished in Japan during the centuries of isolation from Western influences.

In the fall of 1941, a few months after I met Pedoe, I went as a student to Trinity College at Cambridge University. There I listened to lectures by Hardy and the other famous English mathematicians J. E. Littlewood, William Hodge, and Abram Samoilovitch Besicovitch. My favorite was Besicovitch, a Russian émigré who worked on the boundary between geometry and set theory. At that time, in the middle of the war, there were very few students at Cambridge, and I had Besicovitch all to myself. He taught me Russian as well as mathematics. We went regularly for long walks, on which only Russian was spoken. He had a billiard table in his living room, and we played billiards when the weather was too wet for walking. Besicovitch gave me problems to work on that were impossibly difficult but taught me how to think. His great work on the geometry of plane sets of points became a model for my own later work in physics.

After two years as a student and two years working as a statistician for the Royal Air Force, I had my first academic job at age twenty-one—as instructor in mathematics at London University. I attached myself as an unofficial student to Harold Davenport, a number theorist who was a professor at University College London. So I became a number theorist. Davenport was, like Besicovitch, a friend as well as a mentor. He gave me problems to work on that were easier than Besicovitch's problems: Davenport's problems were difficult, but not impossible. I solved two of them and published the solutions in mathematical journals. Davenport was one of those rare people who can gauge the capacity of a student and supply a problem just hard enough but not too hard for the student to solve. I owe to him my start as a professional mathematician.

While I was working with Davenport, I was also thinking seriously about switching from mathematics to physics. I read the Smyth Report, *Atomic Energy: A General Account of the Development of Methods of Using Atomic Energy for Military Purposes Under the Auspices of the United States Government, 1940–1945*, which was published in the fall of 1945 and described the achievements of the physicists who had built nuclear reactors and bombs during the war. The report gives a vivid and detailed picture both of the science and of the scientists, and I felt strongly tempted to join the club. The mathematics I was doing would never be of interest to anybody outside the small community of number theorists. It was not even modern mathematics; it belonged to the nineteenth rather than the twentieth century. If I wanted to become a modern mathematician, I would have to go back to school and learn the modern stuff. Why not learn physics instead? Physics had two great advantages: First, I would be doing something important rather than solving esoteric puzzles; second, physics was well suited to my skills, since the mathematics needed for physics was nineteenth-century rather than twentieth-century mathematics. For me, the switch to physics was not difficult; it meant only that I would have to find a physicist as helpful as Davenport to supply me with problems to solve.

As a result of my work with Davenport, Trinity College invited me to come back to Cambridge as a research fellow. That meant I was free to do anything I wanted. What I wanted was to leave

England and see the world. After six years of war spent either fighting or cooped up in England, everyone wanted to travel. I applied for a Harkness Fellowship to spend a year in America and was one of the lucky winners. At the Cavendish Laboratory in Cambridge, I happened to meet Sir Geoffrey Taylor, an expert in fluid dynamics who had been at Los Alamos during the war. I did not know Taylor, but I ventured to ask him where I should go in America. He said without any hesitation, "Oh, you should go to Cornell. That is where all the brightest people from Los Alamos went after the war." Hans Bethe, who had been head of the theoretical division at Los Alamos, was the person I should work with, he said. Taylor knew Bethe well and would put in a good word for me. I knew almost nothing about Cornell, but I took Taylor's advice and went to Cornell to work with Bethe. And Bethe turned out to be even better than Davenport. He gave me a difficult but not impossible problem to solve, and I published the solution in a physics journal. That gave me confidence. Now I really belonged to the club.

My final stroke of luck was meeting Richard Feynman, who was then a young professor at Cornell, not yet famous. I had never heard of Feynman before I came to America. He was rebuilding the whole of physics from the bottom up, using a geometrical language, with diagrams that nobody except Feynman understood. I realized that Feynman was a genius and my job was to understand his language and explain it to the world. So that is what I did. I spent as much time as I could with him. I watched him drawing his diagrams on the blackboard and listened to him talking. Like Besicovitch, he used to go for long walks and talk about everything under the sun. After a year at Cornell, I understood Feynman's way of thinking and translated it into the old-fashioned mathematics I had learned in England. I published two papers explaining why Feynman's methods worked. My papers were bestsellers, and Feynman's language became the standard language of particle physicists all over the world. At the age of twenty-five, I was a famous physicist! At a meeting of the American Physical

Society, where I was one of the main speakers, Feynman said to me, "Well, Doc, you're in." Childhood was over—and now I was free to spend the rest of my life finding problems in various areas of science where a tablespoonful of elegant mathematics could make a big difference.

A Strange Beautiful Girl in a Car

LEE SMOLIN

LEE SMOLIN is a founding member and research physicist at the Perimeter Institute for Theoretical Physics in Waterloo, Ontario. A prominent contributor to the subject of quantum gravity, he is also the author of *The Life of the Cosmos* and *Three Roads to Quantum Gravity.*

What one believes as an adult about one's growing up is to a large extent a personal mythology, behind which is the working out of dramas involving one's parents. My parents wanted to be artists. My mother succeeded, becoming a playwright, but not without a struggle. My father wanted to be an architect or a sculptor and instead became an engineer. I don't recall either of them ever telling me what to do, but it was somehow communicated to me at an early age that I was to have a creative life, and that what one accomplished in life creatively was more important than any ordinary measure of success or stability. Perhaps it is as a consequence that I have always shied away from competition, whether it was tennis matches as a teenager, sailboat races as an adult, being the best student in the class, or working in a competitive area like string theory. What has mattered most was doing something creative and lasting. Beyond that, impressing girls was always more important to me than impressing teachers.

One of my earliest memories is of going on walks with my father across Central Park to observe the progress of the construction on Frank Lloyd Wright's Guggenheim Museum. I also recall a few years later our reading together a popular book on relativity theory, and drawing pictures of trains and lanterns. As a New York City child, I was fortunate to go to an elementary school for bright children. Then, when I was in fifth grade, we moved to Cincinnati, in the middle of the school year. My new teacher gave me the math book and asked me to tell her what I had learned. I knew all of it, but for some reason I was afraid to tell her that. I was put briefly in the class for dumb kids in math, then eventually advanced.

At about this time, I told my mother I wanted to figure out what the purpose behind the universe was. She told me that that would make a good topic for a PhD thesis. Another boy and I discussed forming a club in which we would discuss the meaning of the universe. I had some ideas (which now I can't remember) and we debated them. But we never got anyone else to join.

My parents practiced a kind of benign neglect; their strategy was to support almost anything I took an interest in as long as it wasn't too dangerous. I was under no pressure to get good grades, and indeed my grades were erratic. But they supported me in whatever I did. When I had a rock-and-roll band, they put up with rehearsals in the living room and defended me when the neighbors called the police. When I was twelve, my father took me on a drive, and on our return, sitting in the car outside the house, he told me in plain terms what he expected of me. I was not to do drugs, drink excessively, get a girl pregnant, or be homosexual. If I refrained from these things, I'd be fine. I never did any of them, and apart from fights about mowing the lawn, in which I tried to advocate the right of the grass to grow naturally, I conformed to my parents' expectations. (OK, I did smoke marijuana twice—once at fourteen and a second time with a girlfriend at twenty-five. I experienced being outside of my body and looking down from the ceiling at myself, and I hated it.)

I don't recall having much interest in science. In seventh grade I went to a summer enrichment program in science—and I played with magnets, wires, chemicals, all the usual stuff—but I don't recall any of this making much of an impression. I have never in my life taken anything apart to fix it or to see how it works. I became a scientist after all because of two mentors who made everything possible. The first was a friend of the family, William Larkin, a mathematician at Xavier University, who let me play on his department's computer at a time when computers filled large rooms and no one thought a ten-year-old could write a program. As a result, I was invited to give a talk at a national teachers' conference in downtown Cincinnati. A newspaper photo shows me standing on a platform in a little suit, telling the teachers how to write a simple program in FORTRAN.

Bill Larkin and I had an understanding that in fact I was not special at all—that my case was meant to demonstrate that a typical ten-year-old could program a computer. But my feelings were hurt when I tried to write a program to solve a puzzle and couldn't understand why it didn't work. The puzzle was to draw a certain figure without either lifting one's pencil from the paper or retracing any line. I showed it to a college student working in the computer center. He laughed at me and explained how to prove, using an idea from topology, that there was no solution—a possibility it had not occurred to me to include in my program. To this day, I don't like solving puzzles.

The greatest gift Bill Larkin gave me was arranging for me to learn calculus early. When I was in ninth grade, my high school informed me that I did not rate highly enough on tests to take the advanced track, leading to college-placement courses, in both mathematics and English. I had to choose one or the other, they said; both would be too challenging for me. I decided to choose English, since I was thinking of becoming either a writer or a musician. My parents must have told Bill Larkin about this, and he came to the house and made a proposition to me. If I would take a precalculus course at his university that summer, in which I would learn the next two years of high school math, and if I did well, he would persuade the high school to let me take calucus the next year.

I accepted his challenge out of a sense of rebelliousness against the high school and its reliance on testing. I was not very interested in mathematics, but to prove them wrong I worked hard and succeeded. I was a rebel. The previous year, I had helped organize a visit by Jerry Rubin—the Yippie, antiwar activist, and professional provocateur—back to Cincinnati. Rubin had gone to my high school and wanted to make a triumphant return as background for a story that *Look* magazine was writing about him. With a few friends, I organized a concert in a local park, where he spoke to several thousand people. *Look* referred to us as "the cream of the alienated." As a consequence of this, I never took physics in high school. The physics teacher, who was a passionate man very right-wing in his politics, told me that it had been a mistake to let me learn calculus early and that no matter how well I knew the math, he would not accept me into his physics class.

One morning that year, my father read in the newspaper that Buckminster Fuller, the great visionary architect and inventor of the geodesic dome, who was then in his seventies, would be speaking at a downtown conference. He suggested I invite Buckminster Fuller to give a talk at my high school. I called the hotel where the conference was held and left a message, then went to school. Shortly I was called to the principal's office. When I got there, he was holding the phone with his hand over the receiver. "There's some guy on the phone who says his name is Fuller and you should pick him up at noon, as he's giving a talk here," the principal said. "Who is he? He isn't another Jerry Rubin, I hope." I got my English teacher to tell the principal who Buckminster Fuller was, after which an assembly was called for the whole school. A friend and I drove downtown and picked Fuller up. On the drive, he showed us that he wore three watches, one set for the time zone he was in yesterday, one set for the current day, and one for the time zone he would be in tomorrow. After he was introduced to the assembled high school, Fuller went backstage, got a folding chair, placed it downstage center, put a microphone in each breast pocket, took out his hearing aids, sat down, closed his eyes, and spoke for seven hours. By the end, there were only perhaps a dozen of us there, but I knew I wanted to be an architect.

Soon my room filled up with models of geodesic domes and other exotic structures. That summer I advertised my services as a manufacturer of geodesic-dome swimming-pool covers. Luckily no one bought one, as I had no idea how to put a real one together. But wondering how to ensure that they would not fall down on potential customers, I asked Bill Larkin how one did structural calculations. He said there was something called tensor calculus. Luckily, my study of mathematics had enabled me to read the books in that area. I then wondered if one could make a geodesic dome based on shapes other than a sphere. Bill Larkin let me use the computer to write programs to calculate how to build a geodesic dome based on any curved surface. Even now, the diagrams of quantum spacetimes in my research notebooks look a lot like geodesic domes.

That year I had been attending meetings of a group of teachers who wanted to found an alternative high school. I was enthusiastic, and my parents let me transfer there for my senior year. The new experimental school happened to be across the street from my girlfriend's apartment, which also pleased me. In my first week there, the teachers explained their philosophy: One should go wherever in the community the knowledge one wanted was located. I thought about this for a minute and realized that the knowledge I wanted was located in a university. So I started attending classes at the university where my mother taught. I took courses in literature and advanced mathematics. I soon realized that it was silly for my parents to pay money to the high school while I was attending classes at the university (which I attended free, as my mother was a professor), so I dropped out of high school.

In eleventh grade I had applied to Hampshire College, in Amherst, Massachusetts, a newly founded school where students could design their own programs, and I had been turned down. I now applied again and was told that they wouldn't read my application because I had already been rejected. I called and insisted that they reconsider, as Hampshire was the only place I wanted to go to college. They relented and interviewed me. On the basis of my passion for architecture, and despite having dropped out of high school, I was admitted.

Later that spring my longtime girlfriend broke up with me. I fell in love with a new girl, who lived in the neighborhood. One warm spring evening I took a walk to her house but was told that she was out with some friends. I returned home and picked up a book by Einstein that I had just taken out of the public library. I was curious about Einstein: The math I needed to design buildings with curved surfaces was exactly what he had used to describe the curvature of space and time. That evening, sitting on the porch, I read his essay "Autobiographical Notes." I read for a while and then I strolled around the neighborhood with the book, sitting on the sidewalk a few times under a streetlight to reread a passage. I was hoping to run into that girl. I didn't see her, but in the meantime I came to a decision that my life would be dedicated to following the path of Einstein. One of his ideas that appealed to me was that by becoming a scientist you could transcend the pain and uncertainty of ordinary life. By grasping the laws of nature, you connected with an aspect of the world more permanent and beautiful than the short striving of human life. But I also understood somehow that I could do physics. I knew I wasn't meant to be a mathematician; I was discovering in my advanced math courses that there were people more mathematically talented than I. But I knew I could very quickly learn enough about a piece of mathematics to use it for some end. I had never met a physicist, I had never taken a physics course, but somehow I understood from reading Einstein that this was something I could do. In the essay, Einstein wrote that there remained two big unsolved problems: what quantum mechanics meant, and the relationship between quantum mechanics and general relativity. I decided that evening that I would work on those problems. Indeed, I have worked on them ever since.

The next day I told my parents I was going to be a physicist instead of an architect and I began thinking about how to do it. The spring term at the University of Cincinnati was starting, and I enrolled in a graduate course in general relativity—my first physics course. I also ordered a catalog from MIT and from it wrote out a list of courses and books one needed to study to become a theoretical physicist. I applied to MIT, but I had a backup plan, which was just to study the books in the undergraduate courses on my own and then apply to graduate school. I figured it would take me two years and save a lot of money on college tuition. After a while, I got back together with my girlfriend. She had applied to transfer to Hampshire, so we drove up to Massachusetts for her interview, after which we were to go to MIT for my interview. While she was being interviewed, I walked over to the science building to see if there were any physicists there. They had just hired a young professor named Herbert Bernstein. I told him of my interest and he asked me to explain general relativity to him, pretending he didn't know it. We talked for several hours, and I realized that he was someone I could learn from. I never made the trip to MIT. When I got home, there was a long handwritten letter from Professor Bernstein posing a number of questions about relativity theory.

So the next fall I went to Hampshire College and began studying under Herbert Bernstein. Without him, I would never have become a scientist. He shamed me into doing the hard work necessary to be able not just to talk about math and physics but to calculate. Without that discipline, my story would have been very different; perhaps I would have become one of those lost souls wandering the basement of MIT playing with computers and hacking the telephone network.

Herb made me the grader in the course I was to take with him. This meant that I had to do all the problems ahead of the other students and turn them in to him. After he graded my efforts, I was to grade the others. He was very hard on me; he would call me at 2:00 A.M. and yell at me, using a lot of profanity, telling me how disappointed he was in my efforts. Then the phone would ring at 7:00 A.M. and he would say, "Meet me in my office in an hour and show me how to do problems 8 and 10." Two years later, when I told him I wanted to go to graduate school, he asked me to solve a problem on the board, after which he told me it was just possible I might be good enough. I never understood why I was the only student he wasn't nice to, the only one he didn't encourage or help. For years afterward, I was terrified of him. Now we are good friends, but it took a while.

When I went to graduate school at Harvard, I was already embarked on my research program to try to answer Einstein's questions about quantum theory and its relationship with relativity. Harvard was a tough, no-nonsense place, but they accommodated me by letting me work on quantum gravity even though no one else there did. (At the time, few did anywhere.) From then on, I have been fortunate in spite of myself. I was warned more than once to give up working on quantum gravity in favor of more trendy areas of particle theory, but though I continued in my own direction I always managed to get good positions. In retrospect I can see that from my early teens to the present the emotional side of my life dominated and I focused more on the drama of my personal relationships and attachments than on my career. But I always worked. And I always did the work I wanted—following the two problems I had taken from Einstein's essay. I never once chose a research problem because it would have been good for my career. It's not that I never cared what other people thought. Like most of us, I wanted approval and recognition. But I wanted to be recognized for my own ideas, not for the quality of my contributions to other people's research programs. Given what I know now about the importance of trends and fashions in determining who gets hired for academic positions, I'm surprised and grateful that my career prospered.

Why, then, did I become a scientist, if I took no pains to succeed at the game of making an academic career? What I believe about myself is that my desire to do science was based on the need to have a picture of the world that was coherent and in which I could understand my place in the universe. For some reason I still can't fathom, in order to be comfortable in the world, I need a belief system that runs coherently all the way from the whole universe down to the existence of, if not myself, at least beings like myself. I know there will be those who see this idea as inherently implausible. But if someone asks me why I have worked all my life on quantum gravity, the answer is because the schism between quantum theory and relativity theory is an obstacle to that coherent picture within which I make sense of my own existence and the uncomfortable fact that it will be brief and limited.

Perhaps this seems mystical. I don't know. In college I knew a girl who had a sign on her door: "No mystics allowed." After a while, she added "except Lee." But I have had only one mystical experience in my whole life. I was seventeen. It was the summer before I went to college, and I was in Los Angeles staying with relatives and working as a sheet-metal apprentice to earn some money. I was also teaching myself some basic physics—mechanics, from the book by Landau and Lifschitz. One evening I took a break from studying and went for a walk out into a deserted field near my aunt's house. I was feeling lonely and wondering how one met girls in Los Angeles. For some reason, I sat down on the ground, and all of a sudden an intense emotion engulfed me, uncalled-for and from nowhere. For what seemed like a long time, I felt deeply and completely connected to the universe around me. I was wondrously happy and calm.

Still with this feeling, I got up and walked out of the field toward home. As I crossed the street a car slowed and stopped. A beautiful girl, perhaps a few years older than me, rolled down her window and asked me what my plans were that night. I was too confused to answer. She tried again, and then stared at me for a minute before driving on. I spent the rest of that summer wondering about her and the experience, regretting my inability to hold on to either. In the years following, I often took walks in woods and fields, hoping for a return of the feeling of oneness with the world, or at least of a strange beautiful girl in a car. To this day, nothing like either experience has ever happened again.

How We May Have Become What We Are

STEVEN PINKER

STEVEN PINKER, an experimental psychologist, is Johnstone Family Professor in the department of psychology at Harvard University, and the author of, among other books, *The Language Instinct, How the Mind Works, Words and Rules,* and, most recently, *The Blank Slate: The Modern Denial of Human Nature.*

Don't believe a word of what you read in this essay on the childhood influences that led me to become a scientist. Don't believe a word of what you read in the other essays, either. One of the curses of being an experimental psychologist is the habit of scrutinizing one's own mental processes. Recounting childhood influences is a mental process no less subject to quirks and errors than falling for the visual illusions on the back of a cereal box. Everything I know about the recollection of childhood influences makes me approach this assignment with misgivings.

My cynicism about the recall of juvenilia began about ten years ago, when an interviewer asked me a question: "The dedication to Stephen Jay Gould's first book reads, 'For my father, who took me to see the Tyrannosaurus when I was five.' What was the equivalent moment in your childhood that led you to devote your life to understanding language?"

I was dumbstruck. The only thought running through my head was what a genius Gould was for coming up with that charming line. The interviewer tried to put me out of my misery: "Was it growing up in Quebec, where language was always in the headlines?" Gratefully, I muttered, "Yes, that's right," knowing all the while that it was a load of codswallop. I acquired my interest in language only in graduate school, and it didn't become the focus of my research until I was a tenured professor. Also, the way that language served as an ethnic badge in Quebec had nothing to do with the parts of language that interest me—namely, the mechanics of grammar and vocabulary. But it was better than anything I could come up with on the spot, and I nodded away. Since then I have prepared a few one-liners to answer the question, and I dispense them as the mood strikes me. Some of them appear in this essay. But I don't believe a word of them.

The fallibility of human memory is one reason. People "remember" everything from watching the Kennedy assassination on live television to being abducted by aliens; the unreliability of memory has become a staple of the psychology curriculum. But it isn't the only reason for my skepticism. Even when we remember events accurately, we are apt to misidentify their place in the causal tapestry of our lives.

In a classic 1977 review, the psychologists Richard Nisbett and Timothy Wilson argued that many of the causes of our choices never enter our consciousness. Here is a simple example. If you present people with an array of articles of clothing and ask them to pick one to keep, they tend to pick the rightmost one. But if you then ask them to list the reasons they chose that article, no one says, "Because it was the one on the right." They cite only the features of the objects themselves. Not having served in experiments in which the same items were presented in different orders, people have no grounds for knowing that a dumb factor like left-to-right position could be a cause of their behavior. And that's a major problem for memories of what influenced us: None of us has taken part in the experiments that would isolate the causes of our choices in life.

The first of these invisible causes is our genome. As the psychologist Hans Eysenck once said, the largest influence parents have on their children is at the moment of conception. But other than someone who was separated at birth from an identical twin, or someone who was adopted with another baby and raised as a "virtual twin," people have no way of detecting the effects of their genes on their choices. A few famous twins, like the advice columnists Ann Landers and Abigail van Buren or the university presidents Bernard and Harold Shapiro, offer a hint that career choices might be genetically influenced. Systematic studies confirm that the influence is real, albeit statistical.

What people do often cite is the influence of their parents. An autobiography, like an Academy Award acceptance speech, is an opportunity to thank the people who have been kind to us. Not to acknowledge one's parents would be the height of filial ingratitude. Yet a second finding from behavioral genetics shows that parents have less of an influence than people think. By most measures of talent and temperament, siblings separated at birth end up no more different from each other than siblings raised together, and adopted siblings raised together end up not similar at all. That means that everything a pair of children in a family share when growing up—parents who are engaged or distant, warm or cold, tidy or disorganized, refined or coarse—has little or no long-term influence on making us who we are.

Behavioral genetic studies offer a third blow to reminiscences about childhood influences. Identical twins reared together share all their genes and most of their environment: their parents, home, school, neighborhood, and peer group. Yet correlations in their personality traits are no greater than fifty percent. Identical twins are thus far more similar to each other than unrelated people, or even ordinary siblings, are to each other, but they are nowhere near being indistinguishable. With heredity and environment pretty much the same but the outcomes far from identical, chance must play an enormous role in development. We might be shaped by whether an axon zigged or zagged as our brains jelled in the womb, whether we got the top bunk or the bottom bunk, whether we were dropped on our head, whether we inhaled a virus. Needless to say, few people cite factors like these among their childhood influences.

With constitutional factors (genes and chance) being important but invisible, people tend to blur cause and effect in thinking back on supposedly formative childhood vignettes. One of the contributors to this volume writes endearingly of exploring nature in a little-known stand of forest in the Bronx. Now, one would hardly claim that growing up in the Bronx predisposes a person to a life of exploring nature. More likely, children with a scientific bent seek out nature wherever they can find it. The conventional wisdom might have it backward. Rather than childhood experiences causing us to be who we are, who we are causes our childhood experiences.

It is not only ignorance of the causes of our life path that leads us to confabulate our early influences. It is also the demands of the narrative genre. Accurate renderings of life are famously boring—just think of home movies, vacation slide shows, and reality television. (*The Osbournes* is the exception that proves the rule. Its entertainment value comes from the clash between our expectation that the bad-boy rock star might bite the head off a bat at any moment and the suburban banality of the life he actually leads.) When asked to submit an essay about our lives, we become content providers who edit the events into the satisfying arc of a good plot. Chekhov remarked that if an audience is shown a gun in act 1, they can count on it going off in act 3. In composing a story line for our lives, knowing that a gun has gone off in act 3 tempts us to show it to the audience in act 1.

There is one more reason that autobiographies are less than truthful: We all want to look good. Experiments have shown that most people explain their behavior in ways that put them in the best light. We all sell ourselves as capable, noble, consistent, and in control of our lives, so that we can entice people to befriend us, trust us, or empower us. This explains the well-known phenomenon of cognitive dissonance reduction, in which people say they enjoyed a boring task for which they were paid a trifling sum because they don't want to admit that they can be manipulated by social pressure. Likewise, many participants in the famous Milgram obedience experiments said that the victims deserved the punishment, because no one wants to look like either a sadist or a lackey. I once read an interview with the legal scholar Susan Estrich in which she explained why defense lawyers insist that their clients not talk to the police. It's not that guilty clients will let slip the truth; it's that innocent clients will try to look good, and in doing so will talk themselves into a contradiction that the prosecution can use to impugn their credibility.

We all maintain self-serving theories of how our lives unfolded, and the social psychologist Michael Ross has made a strong case that people's recollections are tainted by these theories. Memory is Orwellian, constantly rewriting the past to conform to present exigencies. If you give people a persuasive argument that vigorous tooth-brushing is unhealthy, most of them not only change their minds but deny that they ever believed otherwise, and readjust estimates of how vigorously they brushed in the past. If you put people in a worthless program to improve their study skills, most of them retrospectively down-rate their previous study skills, since they want to think they've improved and it's harder to falsify the present. Most people subscribe to the false theory that they become better adjusted as they age, and as a result they misremember their own ratings of adjustment made twenty-five years earlier as being more negative than they were. In the other direction, most people overestimate how rapidly memory declines with age. Once again, they conform their recollections to their tacit theories, and in their sixties they overestimate the memory skills they enjoyed in their thirties.

So why do people *really* go into science? I suspect the real stories run something like this. Thanks to genes and chance, some people are born with a dose of the requisite talent and temperament:

curiosity about the natural world, mechanical and mathematical aptitude, a tilt toward intellectual compared to physical and social forms of amusement. Unless their childhoods were unusually deprived, they will have been exposed to the cafeteria of opportunities provided by modern societies and will have gravitated to a succession of teachers, activities, books, clubs, courses, friends, and hobbies that engage them. For this reason, I am not claiming that the environment is unimportant in shaping us. It's just that the relevant environment—culture and society—is pretty much the same for everyone in the group of people to whom we implicitly compare ourselves, and thus plays little part in the story of how we became the individuals we are.

In most cases, people reach adolescence with no clear idea that they want to become a scientist or anything else, though they may flit among a set of plausible careers that are familiar to them. (How many children are even *aware* of the professions that will become their life's calling: mycologist, actuary, endodontist, comptroller, mortgage-loan specialist?) At some point in college, they find themselves with a course or friend or work-study job that exposes them to a way of life that feels congenial. The goals excite them; the daily grind is pleasant; the peers share their interests, values, and sense of humor; the pay, hours, and prestige are acceptable. They increasingly specialize in that field through graduate training and beyond—and there they are. It doesn't make an interesting story, but I suspect it is an accurate one.

I like a good story as much as anyone, and here is one that pleases me. I was born into the Jewish community of Montreal—a minority within the English-speaking minority—which numbered about a hundred thousand at its peak and claims such figures as Leonard Cohen, William Shatner, Moshe Safdie, Edgar Bronfman Jr., Mort Zuckerman, Charles Krauthammer, Saul Bellow, and Mordecai Richler. The community was a generation closer to Europe than its American counterparts, because the United States closed its door to most immigration from the 1920s to the 1960s and Canada was the next best choice for Jews fleeing

Europe, including refugees from the Holocaust, Stalin, and the Hungarian Revolution. (My own grandparents emigrated from Poland and Bessarabia in the 1920s.) I remember it as an argumentative and intellectual milieu (though college was a luxury in my parents' generation), like 1930s New York. We had a saying, "Ten Jews; eleven political parties." Montreal was also fertile ground for an interest in psychology. D. O. Hebb of the McGill University psychology department and Wilder Penfield of the Montreal Neurological Institute were celebrities; a boulevard in the city was later named after Penfield. Both led influential programs in their institutions, and an unusual number of cognitive psychologists and neuroscientists emerged from this milieu.

I chose my parents wisely. Both are talented in language and math, my mother fascinated by ideas, my father by gadgets. As the oldest grandchild of immigrants, it was a foregone conclusion that I would become a professional; failure was not an option. Our house had books, magazines, and the *World Book Encyclopedia,* which I read in its entirety. I read biographies of scientists like Pasteur, Lister, Banting, and Einstein, as well as George Gamow's delightful *One, Two, Three, Infinity,* which I consult to this day. We subscribed to a monthly series of Time-Life books on science, each with a different-colored spine naming its topic: *Electricity and Magnetism, The Planets, Evolution, Light and Sound, The Earth,* and a really interesting one called *The Mind.* I still remember many of its contents, including Freud, visual illusions, IQ, sensory deprivation, maze learning, and schizophrenia. When I first showed an interest in the mind, my mother suggested that I become a psychiatrist, because psychology, as it was taught to her in the 1950s, was about "cats and rats."

My education was, by today's standards, mediocre. Quebec's bizarre school system divided children by language and religion, and all non-Catholics were assigned to "Protestant" schools that squandered half an hour a day on Christian hymns and Bible stories. I learned to read from Dick and Jane primers, based on the now discredited look-say method, and was subjected to fads such as New Math and Skinnerian programmed learning. The science education was Victorian. Canada had not been galvanized by *Sputnik*, and in those prefeminist days the teaching profession was a pink-collar ghetto peopled by young women who disliked science. I learned little more than that beavers build dams, iron rusts, warm things expand, and cold things contract. When I challenged a teacher by noting that milk bottles left on our porch on a winter day had burst, she said it could not have happened, because warm things expand and cold things contract.

Schooling is a small part of education, and I learned science anyway. There were the books at home, and science in the news, particularly the Mercury, Gemini, and Apollo space programs, which became an obsession. Short circuits in a train set taught me about electricity. A failed science project, based on a suggestion in a library book, taught me about chemistry: A coil of wire attached to a battery and dipped in an electrolyte solution did not act like an electromagnet, but the tip of the wire got thickly coated in copper, and I soon amazed my friends by producing copper-plated dimes and nickel-plated pennies.

The psychologist Judith Rich Harris has persuaded me that peers are paramount in socializing children, and I realize that the most profound influence of my schooling was to put me together with intellectually engaged peers. A short-lived program (unthinkable today) gave every student in the district an IQ test and skimmed the highest scorers into a two-year enriched class at a central school. In that class, I became friends with two extraordinary fellow twelve-year-olds who had an interest in science and much else: Steven Sigler was a red diaper baby, whose father was a Communist and whose older brother was a student radical. We had endless discussions of Marx, Bakunin, and Krapotkin, which temporarily converted me to anarchism and permanently interested me in questions about politics and human nature. (Anarchism, for example, is tenable only if you believe that humans are naturally cooperative and peaceable.) Brian Leber was a preteen polymath with a keen interest in scientific psychology. He exposed me to Hans Eysenck's popular book *Fact and Fiction in Psychology* and dragged me to a public lecture by Hebb at McGill, of which I understood not a word.

I also went to Hebrew and Sunday school, and then Jewish youth groups and summer camps. Hebrew grammar is complex but mathematically beautiful, and I remember deducing, to my astonishment and delight, that *lifnei* ("before") literally means "in the face of" and can lawfully be derived from *panim* ("face"). It may not be a coincidence that Noam Chomsky, the son of a Hebrew philologist, did his first technical work in linguistics on the morphophonemics of modern Hebrew. The language begs to be analyzed in terms of deep structures and cascades of rules, and perhaps this awakened a taste in me for that style of analysis, as it surely did in Chomsky.

My Sunday school, fortunately, was not a conduit for perpetuating dogma but an arena for debating ideas. I remember challenging a teacher who said that the Torah had to be inscribed on parchment with vegetable ink. If the ideas were what mattered, I asked, why couldn't the Torah be transmitted on IBM punch cards? This was, perhaps, the earliest hint of my conviction that information, rather than the physical substrate carrying the information, is the key to understanding mental life.

These are a few childhood events that might be antecedents of my adult passions for language, cognition, experimental psychology, human nature, and the world of ideas and argumentation. Truth be told, the formative experiences probably came later, but as far as early influences are concerned, this is my story and I'm sticking to it.

Patterns and the Participant Observer

MARY CATHERINE BATESON

Mary Catherine Bateson is president of the Institute for Intercultural Studies, in New York City, and professor emerita at George Mason University, where she held the Clarence J. Robinson Chair in Anthropology and English. She is currently visiting professor at the Harvard Graduate School of Education. Bateson is the author of numerous books, including *Composing a Life; Full Circles, Overlapping Lives: Culture and Generation in Transition;* and, most recently, *Willing to Learn: Passages of Personal Discovery.*

An only child, both parents scientists. There, already, is half the story.

In large families, children create their own worlds and find amusement and stimulation in each other. If you are an only child, you spend a great deal of time listening to what the grown-ups are interested in, getting the flavor early for lack of something more immediately appealing, especially if you are encouraged to ask about whatever you don't understand. It also makes a difference if your parents are colleagues. In many households, the children of my generation heard little conversation about work and ideas, because so often the father was engaged in work to which his wife had no access. Scientists left their work in the lab or the office, rarely bringing it into the private sphere. Neither of my parents—the anthropologists Gregory Bateson and Margaret Mead—

made much division between their professional and personal lives. Theories and observations filled the conversations I listened to at breakfast, lunch, and dinner. When my father was away, and after my parents separated, the same pattern continued with whatever colleagues and friends came visiting. Breakfast, lunch, and dinner.

When I was growing up, it was supposed to be fairly clear that boys modeled themselves on their fathers and girls modeled themselves on their mothers, but following the same-sex pattern often involved competition and rebellion, especially for boys, and limited the options available to girls. I had the still unusual experience of growing up in an egalitarian household, in which my two parents were strikingly different and both available as models, with no gender rules determining the choice. In fact, one of the things my mother drew from her field experience to implement in my upbringing was the notion that children need close contact with a number of adults who can serve as alternative models for life choices, ranging from business to the arts and from science to homemaking. Not that this was an entirely representative spectrum, however: I grew up convinced that the normal course of things was to go to graduate school, and that most people wrote books. I always replied, when asked the standard question, that I was going to be a scientist when I grew up—not an anthropologist, but a scientist—and given a choice of courses or activities, I went for science and math.

Both my parents had come to anthropology from other fields, Margaret from psychology and Gregory from biology. Both were involved in the efforts of that era to promote interdisciplinary thinking. They were members of the postwar Macy Conferences on Cybernetics, searching for models connecting the human sciences to other sciences and to engineering, and they were involved with new ways of forging cooperation under the rubrics of behavioral science or human relations or new fields like child development. Once, when I asked my mother what kind of scientist I might end up being, she suggested—rashly, because I teased her about it for years—that I might become an embryologist or a crystallographer. I believe now that what she meant was not that I would be interested in embryos or crystals but that I would be interested in thinking about pattern in a fairly abstract way. She was not commenting on the *what* of science but on the *how,* and not so much the *how* of investigation and experiment as the *how* of intellectual analysis. "Pattern" was an important word in our household, and the ability to observe and describe patterns was key, but for Margaret the patterns of human behavior were primary, so her answer was, I believe, an indication that she felt I was intellectually more similar to my father than to her.

In fact, I didn't realize until I was in my teens that my father was concerned with human behavior. His idea of spending time with me echoed his own childhood activities as the youngest of three brothers. His father was a distinguished geneticist who played a key role in the acceptance of Mendel's work and, in fact, coined the word "genetics." When I think of Gregory, I think of studying tide pools, collecting beetles, constructing an aquarium, and taking and developing photographs together, but also of logical puzzles and problem solving. He explained Mendelian ratios and diagrammed the different kinds of electrical circuits as we searched for dead Christmas tree bulbs in old-fashioned strings wired in series. His rare letters to me over the years contained little of events or feelings. Instead, they were full of diagrams: the legs of beetles, bubble nests built by fish, the emergence of buds on plants. I would later approach all of mathematics and physical science, and eventually linguistics, in terms of his kind of thinking, taking pleasure in the analysis of pattern and organization.

Gregory shifted the *what* of his work fairly frequently: ritual in New Guinea, child rearing in Bali, schizophrenia, family structure, alcoholism, dolphin communications, octopuses. He was an observer and theoretician rather than an experimentalist, focusing on patterns of thought and communication and their possible distortions, devising ways of capturing and comparing naturally occurring behavior. When Gregory set out to photograph a Balinese cockfight, he left out the most photogenic part, the dramatic savagery of the birds in combat, and focused on the identification of the handlers with their birds, as shown by their hands moving in reflection. Later he became a pioneer in the analysis of interactions among family members in filmed and recorded psychiatric sessions.

By the time I was seven, Gregory had moved out, first to Staten Island and then to California. Our visits were spent on natural history field trips. The goal was to see and, if possible, to photograph animals. He was quick at catching snakes or reaching up under the bark of a dead tree and finding a bat. On camping trips in the Sierras, we dragged lures made of bacon and fish along trails in the forest and then sat up most of the night in his car to see what creatures would show up, with a trip wire for the flash camera. We sat in blinds on the foggy California coast to photograph water birds. Because these activities involved a fair amount of patient waiting, he would explain bits of biology to me—like how amoebas divide—or offer me classical paradoxes and mathematical puzzles to wrestle with.

He spoke about insects or plants as well, but like any child who does better at thinking about seals and whales than about plankton, I paid the most attention to the vertebrates, for children focus best on what moves in front of them and what they can identify with. The basic problem in teaching ecology is that the patterns to be understood (and protected) include much that is slow or invisible: carbon and nitrogen cycles, currents of water and air, the microscopic life in the soil. Gregory introduced me to ecology by helping me set up an aquarium, which required thinking about plants and wastes, about the effects of sunlight and the natural maintenance of clear water. Most people today filter and aerate their aquariums artificially, which increases the number of fish that can be kept in a limited space and bypasses the need for living plants. (Often the plants are plastic.) What most tropical-fish fanciers treasure is the beauty of exotic species, vividly flickering or floating dreamily through the water; what interested Gregory was the organization, the pattern, and that was what he spoke about. The challenge he posed was the cybernetic one of maintaining the aquarium in balance, with my own role as one of many interacting factors in its ecology, and the deeper challenge of empathy with such a system.

My mother had a different agenda: to cultivate my appreciation of the variety and potential of human life. She made a sustained effort to expose me to cultural differences: different races, different religious services, visitors from all over the world where she or her colleagues had done research—like meeting the first troop of Balinese dancers to come to New York City or Native American performers in Madison Square Garden's annual rodeo. Most of Margaret's prewar research had been on child-rearing patterns in the six South Pacific cultures she had studied, so she was intensely aware of what was happening in my development and among my peers. Our conversations trained me in the habit of reflecting on experience. In effect, she taught me to be a participant observer of my own life, looking for patterns and for the ways in which meaning was changed by context. She taught me to notice how the rules differed in different households I visited, how to adapt to them, and how to think about and understand the puzzling reactions of adults. Not everyone was upset by profanity, for instance, but some were, and she taught me the importance of controlling my own use of words and understanding different contexts so they would not slip out inappropriately. I remember once in a household where I visited often that I had gotten into a shouting argument (part of the style of that household) in which I called the mother a "witch," without great effect. Then, in a moment of clarity and curiosity, it occurred to me that a tiny, basically arbitrary linguistic change might make a difference in the interaction, so I quite deliberately and experimentally called her a "bitch"—and the roof blew off! Fascinating! I had no name for what I had discovered, until my freshman year of college when I encountered the basics of phonology.

Margaret frequently put things in ethical terms. An understanding of behavioral patterns and systems of meaning was important in building better relationships and effective communication. A friend who sometimes looked after me had lost her peripheral vision due to a brain tumor, which made walking in the city with small children particularly nerve-racking, so my mother explained the nature of her partial blindness to me, pointing out that it was unfair to exploit it. There were refugee children in my school, struggling to learn English and adapt to a new country, who were likely to be teased and to react with blows when words failed them, so we discussed what they might be going through and possible strategies to reach out to them. From her I learned to think of adults not just as authorities and/or resources but as individuals responding out of their own backgrounds and personalities. As a six-year-old, I cut my foot walking barefoot and my companion asked if I wanted to return home or go on to the household of Lawrence Frank, a social scientist, where I was equally at home. I made my decision: "Daddy knows about nature, but Uncle Larry is better with wounds."

Participant observation is the most basic methodology of the cultural anthropologist. It depends on the ability to perceive patterns while still carrying on with daily life. My father was a highly selective observer, generally careless about keeping records, while my mother was greedy for detail. Unlike most ethnographers, she regarded the notes of what she had seen and heard in the field, carefully typed up and contextualized, as the most important product of her work, for a book is the ethnographer's interpretation, while the notes and photographs are the closest later generations can come to primary data, the nearest equivalent in anthropology to experimental replication. They had very different styles of participation as well. My father's ethical sense had to do with recognizing and respecting patterns of organization, attempting to change them only with great reluctance and only when something had clearly gone wrong. My mother was much more of an activist. Her proposals for change were always based on observation and comparison, however, and she was scornful of political agendas derived from ideology rather than observation. By example and by the questions she asked about my experiences and commitments, it was she who taught me the lifelong habit of participant observation, which has made me a social scientist and writer.

My favorite anecdote of what it meant to be an observer of my own life as well as a participant occurred in second or third grade, when my mother volunteered me for a "play date" with a boy my age who was getting into difficulties at school. She warned me on the way over that he was hard to get along with and that his parents were worried. Afterward, she asked me how things had gone, and I asked her to wait until we got home so I could dictate to her a description of the problems I had encountered, "so if any other child ever has to play with him they'll know what to expect." I was already used to the idea that observation followed by reflection would be valuable not only to me but to other people. When I recall my contributions to my parents' thinking, it strikes me that my father especially valued the way I asked questions that challenged him to clarify some theoretical concept he was struggling with, while my mother asked me the questions, valuing what I could tell her about the various worlds I moved in.

In 1956, when I was sixteen, I went with my mother to Israel, where she was lecturing on the assimilation of immigrants and consulting with authorities. The basic premise was a shared Jewish identity, although the immigrants varied widely in cultural backgrounds, in beliefs, and even in physical types. It was on that trip to Israel that I discovered the *what* of my curiosity, matching it to the *how* I had learned from my parents. I was fed up with my high school, disdainful of the American teenage lifestyle, and loud in my complaints about superficiality and conformity—all of which, I imagine now, must have been an annoyance to my mother, a teenage version of the ideological alienation she found so tiresome. After two weeks in Israel, I proposed staying on, learning Hebrew, entering school to prepare for the national matriculation examinations, and applying to college from there—and my mother agreed. What had gripped me was finding a group of people filled with idealism and excitement about building a new country. The result was that I became a participant observer, struggling to understand an unfamiliar culture, not unlike an ethnographer.

There are plenty of fair, blue-eyed Jews, but not many are named Mary Catherine. As a gentile teenager throwing herself passionately into learning the language—going at it gangbusters and improving every day—I was anomalous and exotic, and the people I met responded with warmth to my curiosity. I had questions about everything from the socialist youth movements to the ingredients of lunchtime sandwiches, and above all the specifically Jewish portions of the school curriculum—Hebrew literature, Jewish history, and the Bible. I was captured by the chance to do what I had been prepared to do by all the emphasis on cultural diversity and attention to patterns in my growing up.

Learning Hebrew was hugely exciting intellectually and led me into Arabic (which is similar to Hebrew but more so), Middle Eastern studies, and linguistics. Young people were once taught Latin "to teach them to think," but the Semitic languages teach a different style of thinking from Latin, one that may encourage an awareness of process and relationship separate from particular things or persons. Families of words are built by combining roots (sets of consonants, unpronouncable by themselves) and patterns (consisting of vowels and affixes) so that, for example, a root with the general meaning of "joining" is brought to life in words for acts like "fastening," "uniting," and "adhering," and for "friend," "society," "alliance," "composition," "notebook," and more, each connected both to the root and to the kind of process or relationship expressed by a given pattern. Roots and patterns interlock neatly, like the fingers of a pair of clasped hands. It's like plugging specific values of *x*, *y*, and *z* into a formula that expresses a particular process or relationship; no wonder Arabic-speakers invented algebra! Semitic grammars are also reminiscent of cybernetics as I encountered it in my parents' work, an analytical system in which organizational (pattern) similarities can be traced from context to context, so that an ecosystem like a forest can be compared to a university or a nation—or an aquarium. Learning Hebrew was exciting to me in the same way as hearing about Mendelian ratios, and called for similar intellectual skills, but it was exciting in another way as well. A new language allows you to think new thoughts. As I learned Hebrew I became convinced that I was not only learning a different way of viewing the world but learning the flexibility to vary my way of viewing the world, to move from concept to concept.

Nevertheless, I would not have become fascinated by Hebrew without the human context, and my year in Israel turned me away from the natural to the social sciences. A decade after my adolescent sense of revelation, with a doctoral degree in linguistics and Middle Eastern studies, I realized that linguistics in the Chomskian era was moving toward the study of pure pattern, and that without the human contact it was not what I wanted to study. I would redefine myself as a cultural anthropologist—in a return toward my parents' profession. My father had known almost no linguistics, but I had learned from him how to think about pattern; my mother had known almost nothing about the Middle East, but I had learned from her how to engage with others reflectively, a participant and an observer at the same time. That began the story of my adult work.

[illegible] values of [illegible] that expressions [illegible] [illegible] no wonder [illegible] speakers [illegible] apart [illegible] of the [illegible] encountered during [illegible] organization [illegible] contrast [illegible] university [illegible] of an [illegible] Hebrew was [illegible] in the same way [illegible] and called for [illegible] skills [illegible] was [illegible] another [illegible]. A new language allowed me to think new thoughts. As I learned Hebrew [illegible] convinced me I was not only learning a different way of viewing the world [illegible] the world [illegible] someone.

Nevertheless, I would not have [illegible] and in [illegible] from the [illegible] the social sciences [illegible] toward the [illegible] of [illegible] patterns, and that without the human [illegible] would [illegible] toward my [illegible] had known almost no linguistics, but I had learned from him how to think about pat- [illegible] knew nothing about the [illegible] East, but I had learned from her how to engage with other societies [illegible] at the same time. That became the story of my adult work.

Mixing It Up

LYNN MARGULIS

LYNN MARGULIS, an evolutionist, is Distinguished University Professor in the department of geosciences at the University of Massachusetts in Amherst. She is the author of numerous books, including *Symbiotic Planet, Five Kingdoms* (with K. V. Schwartz) and (all with Dorion Sagan) *Microcosmos, What Is Life?, What Is Sex?*, and, most recently, *Acquiring Genomes: A Theory of the Origins of Species.*

To survive my parents' squabbles, even fistfights, punctuated with screams of sexual passion that led to four more babies, all female, I invented a multiplicity of escapes. I hid in the back of my father's black-finned Cadillac, out of sight of my friends. The materialistic trivialities of my beautiful, insecure mother did not thwart my father's soaring ambitions. He stopped practicing law ("Most lawyers are thieves and liars"), bought a construction business, and joined a country club. Then, when I was a teenager, he abandoned social climbing forever and exchanged his nouveau riche life to mix with the music-loving bohemians and early hippies.

From age five, when my family moved to the South Side of Chicago, I would lie in the cool grass patch that extended from the traffic-ridden South Shore Drive to our cracked sidewalk. In that tiny natural greenbelt, in view of the glorious Lake Michigan, I studied the frenetic conformity of ants on the sugar trail in grass and sow bugs hidden under rocks. On my belly on the turf, I plotted my escape from the self-centered social-climbing ambience at home, into nature.

From age ten, summer camps on Wisconsin lakes enchanted me from the moment I walked along their shores. At twelve, I became infatuated with science itself when my seventeen-year-old camp counselor began talking about amoebae. She called them a "weird animal." Boy-crazy as I always was, I asked, "How can you tell males from females?"

"You can't," she said. "It's a single cell. It has no sex."

"Then how does it reproduce?"

"It splits in half," she said.

Splits in half?! How did she know? How could that be? Wouldn't it hurt? But in her answer I suspected right then, right away, that my love of nature could be augmented by inquiry. A sense of being able to control my own destiny came with the paradoxical option that I could both learn about and ignore the utter nonsense that surrounded me back home in Chicago. "You need a microscope to see them," she said. Maybe I can watch the strange, boyfriendless creatures, I thought.

As the eldest product of a marriage bereft of sons, I was perhaps the favorite of my father. Although I disdained my parents' world of gossip and getting ahead, politics and parties, golf and cashmere sweaters—the American ethos in all its would-be glory—I nonetheless adopted my father's go-get-'em attitude. I seemed to have inherited his industriousness, his talkative restless energy—even though, as I would later learn, the capacity for energy use is inherited (in mammals, at least) through the mitochondria, the parts of cells that would become one focus of my study. Those cell parts came from my mother, also fiercely energetic but in a much quieter way.

My glamorous mother, though loving, was a quiet, frustrated housewife. Both my parents smoked cigarettes and drank too liberally, and my father cheated. Tormented by his unsuppressed attraction to other women, she was far more concerned that her friends and relatives would learn he philandered than by the act itself. She had little existence outside of him and wanted none. She wanted him home, but he preferred to party. The result was a domestic mess that resulted in transferring to me much responsibility for my younger sisters. Very early I learned well the lesson she taught me by repetition: If you want something done, do it yourself.

We were not poor—my father owned the three-story apartment building in which we lived—and in many ways, despite the dangerous cityscape of Chicago's South Side, we were advantaged: My father was a conscientious provider and my mother was a fine cook. We had social connections and, at times, even servants. But my parents were so busy with their whirlwind social lives and their travails that we sisters were often left alone. I produced and directed family theatricals, in which I usurped the starring roles. We performed these dramas in the building's dark, pipe-festooned basement, bedsheets as curtains draped over the pipes. I was zealous in my demands on my sisters for promptness in rehearsals. I was passionate, intolerant of small talk, hungry for knowledge, grabby, bossy, precocious. Always the serious student, with a myriad of part-time jobs, my few uncommitted moments were dedicated to poetry and daydreaming, with or without props such as books, mulberry trees, and grasshoppers. Whether diary entry or essay, jingle or dialog, if I failed to write on any given day I suffered a sense of deprivation.

Although I can conceal these solitary and bookish tendencies, I haven't changed much. I grew up, as the cliché says, too fast; I was plunged early into the adult world of responsibility. But I also still enjoy an extended childhood: The love of nature, the interest in the out-of-doors and what lies under the microscope, the curiosity whetted by stimulating discussion has never left my life.

I escaped my parents' home into the enchanted land of learning. Hating the reign of terror at Hyde Park High School, I enrolled in the College at the University of Chicago at age fourteen, as soon as I found out about their policy of equality based on test scores, regardless of creed, race, or age above fourteen. I told the university authorities neither that I had permission from my parents (I didn't) nor that I had not finished the tenth grade. This success enhanced my sense that if one applied oneself, much could be done. My rein-taking attitude frightened my mother but was subliminally encouraged by my father.

At fourteen I was mostly a bookworm. My father, a Polish Jew brought up in Protestant communities in Michigan and New Jersey, developed a passion for Palestine transformed into Israel. His enthusiasm, I suppose, had provided adventure that contrasted with the banality of his busy downtown Chicago law office. Although I never shared his burning interest in the rhetoric, depressing tribalism, and fratricides of Middle Eastern politics, I was grateful when he sent me to Israel the summer after I enrolled at the university, to work on a *moshav shitufi* called Molodeth. On this agricultural collective, I picked grapes, milked sheep, and found out that Americans were labeled by the local Israelis as ignorant, greedy, trivial, and naïve. As an American, I was assumed to be materialistic. This taught me about prejudice. But in fact I hated the materialism of the United States. As I became a serious student I gravitated to the academic and the intellectual. I remember my father's rhyming crack, "She used to go with Zionists, now she goes with scientists."

Two factors converted me to science: the University of Chicago and Carl Sagan, who later became my husband.

The College of the University of Chicago was unique, as was its academic beacon, a course called Natural Science 2. Unlike almost everywhere else to this day in higher education, tests were optional. So was all attendance in classes and laboratories, which were limited to a maximum of twenty students. What counted were the six- to nine-hour final examinations in June. These rigorous, thoughtful, and demanding exams were held eight months after classes began in October. They were the sole requirement for a final grade. Also unique to Chicago was the lack of textbooks, which were replaced by direct readings of the great scholars and their commentators. In Nat. Sci. 2, this meant reading

Charles Darwin, Gregor Mendel, Hans Spemann, August Weismann, J. B. S. Haldane, Sewall Wright, and Julian Huxley in an attempt to untangle the great question of heredity. How are the generations linked? What is the nature of that which is passed on from one organism to the next? The provocative questions "What is human?" and "What is life?" and the nature of the cosmos were paraded before our impressionable and engaged minds. Scientific tools of lab work and exposure to science's greatest writers were buttressed by philosophical courses on the nature of scientific inquiry.

Genetics fascinated me most. Nearly everyone, even biologists, still assumes that sex and reproduction go hand in hand. As a postdoctoral fellow much later, I was directed to look closely at laboratory cultures of *Euglena gracilis,* a green swimming microbe, to catch them in the sex act. I never did, because they never do it. This type of failure to document common scientific myth put me on notice that something was wrong in the land of genetics as it was then imagined.

Reproduction is not always preceded by sex. Sex may not even generate variation. My camp counselor, in her relative ignorance, hit closer to the truth about the bizarre sex lives (in the amoeba case, the non–sex life) of reproducing microbes than academic assumption. I later discovered that mates of *Stentor,* another microbe, died every time they paired off. Their sex act, in which neither member of the mating pair has any gender, lasted thirty-six hours. Inevitably it was fatal to both parties. I learned that a related microbe, the ciliate *Paramecium,* could endure the removal of a hefty patch of its surface. A characteristic pattern of its whiplashing appendages known as cilia could be removed and grafted onto another part of its body. When these ciliates reproduced next, they all produced offspring that had inherited the microsurgically displaced parental pattern! In the reprinted syllabus articles of Nat. Sci. 2 we heard, in our minds' ears, the voices of Hermann J. Muller defining life as "mutation, reproduction and the reproduction of mutation"; of Vance Tartar dissecting *Stentor;*

of Theodosius Dobzhansky saying, "Nothing in biology makes sense except in the light of evolution"; and of A. H. Sturtevant trying to understand how chromosomes led to adult shapes in fruit flies. The reality of good data and observation always impressed me more than any argument from authority.

The firm consensus was that the nucleus of the cell in all plants and animals was the sole site of heredity, that the chromosomes in the nucleus bore the all-important genes. Yet intriguing clues, such as that given by the grafted *Paramecium* that reproduced the pattern induced by its surgical operation, made some of us suspect that the nucleus just might not be the sole repository of genetic information. The green parts of plant cells, called chloroplasts, all came from other chloroplasts, outside the nucleus. In *Chlamydomonas*, another green swimming microbe, the energy-generating, oxygen-using part of the cell, the mitochondrion, is inherited from the male, and not from the male nucleus but from the male's mitochondrion, outside the nucleus.

No bacterium, it was becoming obvious, has a membrane-bounded nucleus; rather, they have long thin threads of unbounded DNA, often organized into one long circle inside the cell. Although the idea that the nucleus was central became dogma after the discovery of DNA's role in replication and heredity, a curious observer such as I was—encouraged and trained to read the authentic authorities but also to think for herself—was struck by the fascinating exceptions.

I was sixteen when I met Carl Sagan, and he was nearly five years my senior. Tall, handsome in a sort of galooty way, with a shock of brown-black hair, he captivated me. I literally ran into him one day as I was bounding up the steps of Eckhart Hall, the math building. "Aren't you Lynn?" he asked. "Aren't you Carl Sagan?" I answered. He invited me to the meeting of the Astronomy Club, of which he was president. He already owned a car, a small green Chevy, in which he offered to pick me up at my parents' home early each morning, even though he hated to get up—a trait common to scientists interested in the stars.

At that time he was a graduate student in physics, poised to launch his stratospheric career. Although I was a mere girl and our attraction was the usual erotic one for those of our age, his love for science was contagious. I caught his passion. From childhood he was distracted by ambition and endowed with a self-confidence and love of knowledge that inspired me when it did not run ramshackle over my feelings. As he speculated about life on other planets, and the means to communicate with extraterrestrials, I turned my attention to the humbler abode of Earth and microbes. Genetics, it always seemed to me, gave the best clues of how evolution really works. With that principle in mind, I accompanied Carl north to Wisconsin on the eve of *Sputnik*, September 1957. There, as an astronomy graduate student, he worked at the University of Chicago's Williams Bay observatory, near Lake Geneva. A master's degree candidate, I sought a life sciences education seventy miles away, at a place ideally suited for it: the University of Wisconsin at Madison. The discovery that the Russians had preceded the United States in launching a space satellite, and the political fear this evoked, was to create a climate friendly to scientific funding and research. Both of us young scientists joined the scientific excitement of the nation and the times. Our interests overlapped in questions of the origins of life (a process both cosmic and microbiological) and the atmospheric gases of the planets, which in the case of Earth, as later it became clear, were produced or removed by myriad microscopic and other life-forms.

At Madison, I studied genetics and then population genetics with the master teacher James F. Crow. I adored the former course but felt that the latter, with its zealous concentration on neo-Darwinian concepts such as "fitness," "mutational load," and "coefficients of selection," did not adequately describe the ways in which real populations of live organisms interacted and evolved. Preferring to peer directly into the animated water of living cells, I continued to be intrigued with organelles outside the nucleus. The mitochondria and plastids reproduced on their own, without the nucleus, and they divided neither during mitosis nor by mitosis. Rather, they reproduced by "binary" fission, like bacteria. Although the cell was considered a unit of the animal and plant body—a single being—observations showed that nucleated cells contained more than just nuclei. They were inhabited by bacterial-sized entities capable of replication on their own timetable. Mitochondria and chloroplasts exhibit a sort of rebellion from the rest of the cell. Were they in fact somehow separate, part of a different regime of inheritance? It turned out they were.

My interests in the margins of the cell were complemented by reading interests in the margins of biology. There I found that my predecessors—some of them, like the American Ivan Wallin, maligned and ignored, and others, such as Konstantin Merezhkovsky, taken seriously but only in the Soviet Union—had previously postulated that the organelles had evolved from bacteria that became trapped in larger cells. Their independent evolutionary origin was behind their stubborn tendency to reproduce out of synchrony with the rest of the cell, and to do so like bacteria.

This was a revelation, one that tied together the many observations of nonnuclear inheritance. The differences between bacterial associates and actual parts of the cell could be vanishingly small. Over the course of evolutionary time, associations that began with bacteria feeding on or invading one another could become permanent. These were examples not of predation or infection but of a kind of mating between and among members of different species, permanently—a microbial fairy tale of living symbiotically ever after.

The fairy tale is true. Each cell in your body resembles an amoeba. But the oxygen-using parts, the mitochondria, derive from bacteria. Michael Gray, professor of biochemistry at Dalhousie University in Halifax, and his colleagues by 1983 collected the information that established beyond the shadow of a doubt the multiple ancestry of the eukaryotic cell. Plant cells, and those of algae like the green swimmers *Chlamydomonas* and *Euglena*, had evolved by an additional bacterial association—the incorporation of blue-green photosynthetic bacteria. These alliances, always in each generation subject to natural selection, do not go against Darwin. Rather, they show the power of interliving—symbiosis in the evolution of life. Arguably the most successful life-forms on the planet are not men and women who build telescopes and scan the stars. Neither are they insects swarming in numbers that dwarf those of mammals, but rather the still greater numbers of former bacteria—now the organelles trapped inside the cells of plants and animals. These former free-living microbes provide mitochondrial and chloroplast energy as they function in oxygen uptake and light-harvesting and multiply outside the cell nucleus.

The ancient naturalists speculated about mixed-up animals—chimeras, mermaids, hippogriffs, and sphinxes—that combined parts of fish, reptiles, birds, and mammals. Global exploration and scientific observation show dragons, centaurs, and their kin to be fantasy. Far more amazing than those imagined creatures are the hybrids of our own bodies. Each of us is a colossus of nanobeasts, a coordinated bestiary with abilities more diverse and precise in the aggregate than any machine. If life ever is, as Carl Sagan fondly hoped, found in space, it seems likely that it, too, will embody such a mixed heritage of symbiotic bastards. And in their encounter with us, we all may embark on a future step of symbiotic evolution.

always in each generation subject to natural selection, [illegible] [illegible] Darwin [illegible] the [illegible] of [illegible] [illegible] [illegible] and the evolution of [illegible] [illegible] [illegible] [illegible] of plants [illegible] [illegible] [illegible] [illegible] [illegible] [illegible] [illegible]. Neither are they [illegible] [illegible] in [illegible] [illegible] [illegible] [illegible] [illegible] [illegible] [illegible] greater number [illegible] [illegible] [illegible] [illegible] [illegible] cells of [illegible] and [illegible] [illegible] [illegible] [illegible] provide [illegible] [illegible] [illegible] [illegible] [illegible] [illegible] [illegible] [illegible] and [illegible] [illegible] [illegible] [illegible] [illegible] the [illegible] nucleus.

The ancient [illegible] speculated about [illegible] [illegible] [illegible] [illegible] [illegible] [illegible]—that combined parts of [illegible] [illegible] [illegible] [illegible] [illegible] and [illegible] [illegible] [illegible] dragons, [illegible], and [illegible] [illegible] [illegible] [illegible] [illegible] [illegible] [illegible] [illegible] creatures are the hybrids of our own bodies. Each of us is a [illegible] of [illegible] a [illegible] that [illegible] and has more [illegible] and [illegible] the [illegible] [illegible] [illegible] [illegible]. Wherever [illegible] of [illegible] evolution.

A Childhood Between Realities

JARON LANIER

JARON LANIER is a computer scientist, composer, and visual artist, probably best known for his work in Virtual Reality, a term he coined. Until recently, he was the lead scientist of the National Tele-immersion Initiative, a coalition of research universities studying advanced applications for Internet 2. His current research interests include real-time remote terascale processing, autostereo methods, and haptics.

If I remember my childhood correctly, I discovered the physical world unusually late in life. My earliest memories are of being consumed by an overpowering subjectivity, with only the barest hint that an external, natural world might exist beyond it.

It continues to surprise me how difficult it can be to convey this state of mind to those who do not immediately recognize it. Imagine you are hiking in the light of the full moon at midnight on a high ridge in New Mexico, looking down on a valley dusted in new snow that appears to fluoresce. Now imagine an exchange between two fellow travelers, one a romantic and the other possessing a dry, analytical temperament. The romantic might say, "Isn't this magical?" while his opposite might say, "Well, the nighttime visibility is unusually good, but it's a little cold." In my childhood I was hyperromantic, unable even to conceive of a pragmatic notion like "visibility," because the experience of "magic" was completely overwhelming, to the near exclusion of everything else. My early experience was of the dominance of flavor over form, of qualia over connections. The shorthand term I use sometimes is "mood"—not the mood in one's mind but the mood of a place.

My parents decided to leave New York City in the 1960s in order to raise me, an only child, in an obscure and harsh place—the point where Texas, New Mexico, and Mexico proper join alongside the Rio Grande. This place—a kind of outback—was barely part of America. It was impoverished, relatively lawless, and of unsurpassed irrelevance to the rest of the country. When I was nine, my life was cleanly divided by one moment in time: my mother's death in an automobile accident. She was Viennese and had survived the Holocaust, but not the odd trade-offs we make in order to enjoy American comfort and convenience.

Before her death, I crossed the international border every morning to go to elementary school in Juárez, Mexico. Our schoolbooks were sheathed in fantastic images of Aztec mythology. Through sulfurous memory vapors, I can recall the discovery of a locality that, for the first time, felt fully real to me. Like the rear wall of the closet in the Narnia stories of C. S. Lewis, this place opened up into its hugeness through a small, hidden opening. In this case, the portal was in a ragged old art book on a low shelf in a forlorn school building, and the place was Hieronymus Bosch's triptych *The Garden of Earthly Delights.*

The dusty, mean physical environment around me was not real, but the Garden of Earthly Delights was. Even better was staring at the image while listening to Bach, especially the Toccata and Fugue in D Minor. And even better than *that* was doing both those things while eating chocolate. Amazingly, all these props were available at that Mexican school. It was at this moment that I discovered physicality. The ecstasy was beside the point; what I had found was an instance in which the physical world and the overpowering sense of mood that usually obscured it were connected, instead of being at odds with each other.

Physicality and I had a difficult courtship. My mother's death left me disconnected from the world for a long time. For much of a year, I was in the hospital with a desolate sequence of infectious diseases, barely aware of my surroundings. This awful period came to an end when I experienced two moments of irreversible positivity, brought on by reading just the right sequence of words in my hospital bed. One was the Jewish admonition to "choose life." I was struck by the idea that this adjuration worked on multiple levels. Of course, there was a logic to it, since death would come soon enough, no matter what one chose, so choosing life was at least a reasonable bet. Then the very act of making that calculation implied that you could embrace doubt and that the embrace could be happy. Deeper still was an assertive sensibility that awakened in me: You could make choices! Becoming acquainted with the possibility of will, I had adopted yet another dualistic constituent to my state of mind, to join with subjectivity and its overpowering clouds of mood. Even so, this extraphysical phenomenon was different, in that its only function could be to bring me back to the natural world, where other people would eventually be found.

The second bit of reading was a biography of one of the great early New Orleans wind players who was said to have overcome his childhood respiratory problems by playing the clarinet. I took to playing horns to overcome my own ailments. It worked! I became fascinated by collecting and learning to play new musical instruments, an obsession that continues to this day. My various homes have always resembled forests of musical instruments. They were a way of choosing not just life but mood. Here was a fountain of wonderful subjective experience that was not only correlated to physicality, as with my old magic Bosch portal, but directed by my own body and my own choice.

After release from the hospital, I went back to elementary school, but this time in El Paso. I want to disbelieve my memory of white kids at school bragging about having drowned a Chicano kid in the school's swimming pool, an event that the adult world had registered as an accident. My memories of that school are of a constant onslaught of racism and violence, and of adults who were no better than the children. The thought of connecting with other people—of having friends—was terrifying, and strangers were dangerous. My direct experience was not the only cause of this fear; I suspect that I also inherited some paranoia from my mother, who imported it into our lives from her experience of Nazi Europe.

I didn't know that I should seek to overcome this state of mind, but the odd demographic churnings of the place we lived in eventually brought me into contact with a variety of people, and I slowly learned to connect pleasantly with some of them. For instance, a young soldier at Fort Bliss introduced me to electronics. I found an article about an early electronic musical instrument called a theremin and learned how to build them. The theremin is played by moving your hands in the air near antennae; nothing is touched, and playing it gives you the feeling of contact with a virtual world. (Much later, when I was working with Virtual Reality, I would meet the inventor, Léon Thérémin, then in his nineties; when he learned about my work, he started to vibrate with excitement, like an engine.)

I was also fascinated by the diaphanous luminous images called Lissajous patterns, which can be made by fiddling with musical signals and an oscilloscope, and I made a crude Lissajous viewer out of an old television set. At age eleven, as Halloween approached, a plan formed in my mind: I would build a fantastic haunted house out of my electronic contraptions and attract people worthy of being friends! I hung sheets around our tiny front porch and set up an old enlarger lens to project the Lissajous patterns from the TV onto them. Once the sun went down and the images appeared bright, I felt deliciously surrounded by fantastic dancing forms. The motions of visitors would alter the patterns, as if with the invisible strings of a puppeteer, courtesy of the magical theremin antennae. I wondered whether any girls, those beings of utter mystery, might be delighted by it. Who wouldn't be? My haunted house pleased me immensely but attracted no visitors. Trick-or-treaters steered clear of it. I watched from inside my palace of imagination and freedom as one child after another rejected it, and me. It never occurred to me that they were probably frightened; at the time, they seemed sadistic.

My father and I were struggling financially, since my mother had been the breadwinner, shuffling her investments long-distance, decades before the dot-com, day-trading craze. Eventually the two of us moved to cheap, empty land in southern New Mexico and lived in tents while we gradually built a fanciful home to live in. It took seven years to complete and was composed of spires, crystalline forms, and geodesic domes. My father had unwisely let me design it. You walked in through one side of a cantilevered seven-sided pyramid tilted so that its center of gravity was hanging out over the edge of the foundation. Part of the house eventually fell down, though fortunately just after my father had stepped outside. Alas, I was younger than my friend Lee Smolin when he went through his dome phase, and had no inkling of tensor calculus at the time.

The New Mexico schools were a little less intimidating than those in Texas, but I was no better at connecting with the human beings in them. I was utterly awkward. Coming back into the world after the hospital experience was rough. I simply had no clue what the secret codes were that would elicit any reaction other than contempt. I had to learn interpersonal skills slowly and deliberately. One evening there was a remarkable breakdown of the local telephone system. Anyone who picked up the phone could hear everyone else at once. Hundreds of voices—some sounding distant, some close by—hovered in the first social virtual space I had ever experienced. An instant society of children formed, brilliantly superior to that of the schoolyard, which was straight out of *Lord of the Flies.* The floating children were curious about one another; they were friendly. I was able to communicate with them. The next morning at school, though, no one spoke of what had happened. I looked around and wondered whom I might have talked to the previous night. Was it possible that these rude kids could suddenly become improved, knowable people if the medium that connected us was different?

Most desert dwellers seemed to believe deeply in phenomena invisible to anyone outside their particular group. Besides assorted Native American, evangelical Christian, and Catholic visions, there was the local culture of flying saucers. Kids would bring bits of fallen flying saucers to school for show-and-tell and no one questioned their authenticity, certainly not the teachers. We lived next to the largest missile test range in the world, and peculiar debris fell from the sky all the time. I never believed in flying saucers, but I did find myself entering into the pact of local pride about *our* flying saucers. Beliefs seemed to bring people together, and I hoped in vain that rooting for the home team would bring me acceptance from the other children. I still feel an involuntary surge of indignation when a rival town, Roswell, New Mexico, gets renewed attention for its inferior 1950s flying-saucer crash landing. Our flying-saucer crashes were much better!

My father developed an interest in psychic phenomena, which brought us into friendly contact with other eccentric people and taught me that it was possible not to be rejected out of hand. Some of these psychics were stranger than we were. There was a shaman from the Copper Canyon region of Mexico—home of the mystical traditions that inspired Carlos Castaneda—who had a prosthetic agate eyeball and dressed in ribbons. While the community of paranormal enthusiasts provided me with companionship, I was vaguely ill at ease with them. There was a sense of being exploited. A hot rage would come over me whenever people like the agate-eyed shaman claimed to have contact with my mother—it seemed the ultimate abuse of another person's vulnerability. At least you could trust the sincerity of the murderous kids in the schoolyard. Friendly people could be sneaky. That was a difficult lesson.

There was a social anomaly in our part of the world: a large population of superb engineers employed in the nearby weapons labs, who were mixed into the otherwise undereducated desert population. It was a huge relief to discover the culture of technical people, which was as welcoming to an awkward kid like me as the psychics were but in a way that was not exploitative. One of our near neighbors was a lovely, slight old man named Clyde Tombaugh, who had discovered the planet Pluto in his youth. When I knew him, he directed research in optical sensing at the White Sands Missile Range. He had built marvelous, huge backyard telescopes, and he let me play with them. I will never forget a globular cluster he showed me—a vividly three-dimensional form, a physical object like me, a cousin to me, as real in front of me as anything else in the world. I gained a sense of belonging in the universe. (As an aside, I have no sympathy for the recent campaign to demote Pluto to a prominent Kuiper Belt object instead of a planet. Its weird orbit out there, usually but not always beyond the normal planets, is an inspiration to every kid who doesn't fit in. Let Pluto remain a planet, now and forever!)

When I was fourteen, I enrolled in a local college chemistry class open to high school students and was not entirely forthcoming in filling out the paperwork. I was therefore able to remain in college and never finished high school—a way of escaping from an environment that I feared would kill me. I was electrified by what I learned in college. Still socially awkward, I had the bad habit of stopping strangers in the street to pour my heart out about the latest marvel I had learned. ("Look at these patterns that come out of Abelian groups! Don't you want to see?") But as much as I loved it, there was something missing from the technological world. It was mood, my first and most formative friend. I had encountered fear and loneliness before, but not sterility. I would stay up all night in the math building making hypnotic color patterns with a computer in the dark, feeling like a firefly in a cardboard box. The building was cinder block and barren. The entire world of engineering and mathematics seemed equally barren. I worried that the other nerds, who had by now become my first real friends, might not be able to experience the labyrinth of overpowering moods I had only partially emerged from in order to find some common ground with them. This is the stage of being I remain in today.

I am struck by how many of my adult technological adventures resemble those from my childhood. A Virtual Reality laboratory is something like a grown-up version of my old haunted house. Sensors detect even the most minute body motions and other changes in a person, and exotic devices like 3-D displays immediately generate encompassing sights, sounds, and other sensory events in return. As I worked with Virtual Reality in the 1980s, a hope kindled in me that it might someday—if only in the far future, when I'm long gone—take on some of the qualities not only of my haunted house but also of that long-ago, hospitable telephone-system breakdown. Maybe our descendants will learn to design a virtual agora that brings out the best in people. The Internet has appeared since I grew up, and despite its mixed blessings, so far it encourages my optimism.

What I hope most is that someday all children will be able to build "haunted houses" within Virtual Reality with ease, and that other children will understand them, instead of running away. I believe there is a new form of expression, as fundamental as language, that will come into being as we learn to make virtual worlds a means of interpersonal connection—but that is a topic for another essay. What I will say here is that I hope technology will help children in the future feel that just a little more of what lies deep inside them can be shared than was possible in earlier eras. I observe the rapidly evolving world of social video games expectantly.

I remain troubled by the present sterility of technical culture. The dilemma I face is the same as it was when I emerged from childhood. People who are sympathetic to the world of subjectivity, flavor, and "mood" also tend to break into opposing communities of belief. The agate-eyed shaman was probably a hyperromantic, like me, but he believed in things that excluded me, and even exploited me. The world of engineers and scientists can be competitive and even cruel on occasion, but there's no doubt that all its members live in the same place: the physical world. Anyone and everyone has an equal right to be a scientist or engineer; it's a sad and disempowering mistake to believe that science and technology are solely Western enterprises.

So what is it about science and technology that seems to repel so many people? I'm thinking of the huge numbers who reject Western medicine or subscribe to superstitious beliefs like astrology, not to mention the rise of religious fundamentalism. The most obvious explanation is that people are afraid of death and want to believe in an afterlife, so they are easily exploited by those who offer such messages. But I suspect there's another explanation—one more immediate and harder to dismiss. People who are uncomfortable with a technically adept culture may simply share my sense that subjective experience is fundamental and under threat. When we embrace the natural, or physical, world, we become more egalitarian, since our beliefs don't require that we exclude anyone. But in the process we also deemphasize our subjective experience of the world, which is what validates us as individuals and gives our lives flavor and meaning. And when we celebrate subjectivity, we are in danger of breaking up into opposing communities of belief. In retrospect, I can see how clearly this dilemma contributed to the loneliness I struggled against in my childhood.

My childhood continues.

can be competitive and even cruel no exception, but there [illegible] about [illegible] its members live in the same place, the physical world, [illegible] and [illegible] equal right to be a [illegible] or [illegible] as good and [illegible] to believe that [illegible] any [illegible].

So what [illegible] together going to [illegible] so many people [illegible] of the huge numbers who [illegible] superstitious beliefs, like astrology, not to mention the rise of religion and [illegible]? This [illegible] explanation is that people are afraid [illegible] want to believe in [illegible] easily [illegible] by those who offer such [illegible]. But [illegible] there's another explanation – the more immediate and harder to dismiss, [illegible] with [illegible] subjective experience [illegible] and [illegible]. When we embrace the [illegible] physical world, we become more [illegible] than we [illegible] experience of the world, which [illegible] and [illegible] the [illegible]. And [illegible] I can see now clearly the [illegible] I struggled [illegible] in the childhood.

My childhood [illegible]

Dolittle and Darwin

RICHARD DAWKINS

RICHARD DAWKINS is the Charles Simonyi Professor of the Public Understanding of Science at Oxford University and a Fellow of the Royal Society. His books include *The Selfish Gene, Climbing Mount Improbable, Unweaving the Rainbow,* and, most recently, *A Devil's Chaplain.*

I wish I could say that my early childhood in East Africa turned me on to natural history in general and human evolution in particular. But it wasn't like that. I came to science late. Through books.

My childhood was as near an idyll as you could expect, given that I was sent away to boarding school at seven. I survived that experience as well as the next boy, which means pretty well (some tragic exceptions were lost in the bullied tail of the distribution), and my excellent schooling finally got me into Oxford, that Athens of my riper age.* Home life was genuinely idyllic, first in Kenya, then Nyasaland (Malawi), then England, on the family farm in Oxfordshire. We were not rich, but we weren't poor either. We had no television, but that was only because my parents thought, with some justice, that there were better ways to spend time. And we had books.

Africa and the English countryside should have opened my eyes to the natural world and turned me into a biologist. I had no lack of encouragement from my parents, both of whom knew

*Dryden, of course, in spite of his Cambridge education.

every wildflower you might encounter on a Cornish cliff path or an Alpine meadow, and my father amused my little sister and me by throwing in the Latin names for good measure. But to my lasting regret I showed no aptitude for natural history. I remember my mortification at the age of eight when my tall, handsome grandfather, seeing a blue tit on the feeder outside the window, asked me if I knew what it was. I didn't, and miserably stammered something like "Is it a chaffinch?" Grandfather was scandalized at such ignorance, which, in his outdoors-loving, binoculars-toting, shorts-wearing, Empire-building family, was tantamount to not having heard of Shakespeare: "Good God, John"—I have never forgotten his words, nor my father's sheepish efforts to excuse me—"is that *possible*?" My love of animals came not from watching them, still less from knowing their names, but from books—and not necessarily scientific ones.

I was a secret reader, and it became something of a vice; I would sneak up to my bedroom with a book when I was supposed to be out in the fresh air. Maybe obsessive reading imprints a love of words in a child, and perhaps later assists the craft of writing. In particular, I wonder whether the formative influence that eventually led to my becoming a zoologist might have been a children's book: Hugh Lofting's *The Adventures of Doctor Dolittle,* which I read again and again, along with its numerous sequels. This series of books did not turn me on to science in any direct sense, but Dr. Dolittle was a scientist, the world's greatest naturalist, and a thinker of restless curiosity. Long before either phrase was coined, he was a role model who raised my consciousness.

Dr. Dolittle was an amiable country doctor who turned from human patients to animals. Polynesia, his parrot, taught him to speak the languages of animals, and this single skill provided the plots of nearly a dozen books. Where other books for children (including today's Harry Potter series) profligately invoke the supernatural as a panacea for all difficulties, Hugh Lofting rationed himself to a single alteration of reality, as in science fiction. Dr. Dolittle could speak to animals: From this all else followed. When he was appointed to run the post office of the West African kingdom of Fantippo, he recruited migrant birds into the world's first airmail service; small birds carried a single letter each, storks large parcels. When his ship needed a turn of speed to overtake the wicked slave trader Davy Bones, thousands of gulls gave him a tow—and a child's imagination soared. When he got within range of the slaving vessel, a swallow's keen eyesight aimed his cannon to superhuman accuracy. When a man was framed for murder, Dr. Dolittle, having established his credentials as interpreter by talking to the judge's dog, persuaded the judge to allow the accused's bulldog to take the stand as the only witness to his innocence.

I would argue that there is nothing necessarily supernatural in speaking to animals so that they can understand you, but the animals' doings were frequently mistaken for the supernatural by the doctor's enemies. Cast into an African dungeon to be starved into submission, Dr. Dolittle grew fatter and jollier. Thousands of mice carried in food, one crumb at a time, with water in walnut shells and even fragments of soap so he could wash. His terrified captors naturally put it down to witchcraft, but we, the child readers, were privy to the simple and rational explanation. The same salutary lesson was rammed home again and again through these books. It might look like magic, and the bad guys thought it was magic, but there was a rational explanation.

Many children have power dreams in which a magic spell or a fairy godmother or God himself comes to their aid. My dreams were of talking to animals and mobilizing them against the injustices that humanity (as I thought, under the influence of my animal-loving mother and Dr. Dolittle) inflicted on them. What Dr. Dolittle produced in me was an awareness of what we would now call "speciesism": the automatic assumption that humans deserve special treatment over and above all other animals simply because we are human. Doctrinaire antiabortionists who blow up clinics and murder good doctors turn out on examination to be rank speciesists. An unborn baby is by any reasonable standards less deserving of moral sympathy than an adult cow. The prolifer screams "Murder!" at the abortion doctor and goes home to a steak dinner. No child brought up on Dr. Dolittle could miss the double standard. A child brought up on the Bible most certainly could.

Moral philosophy aside, Dr. Dolittle taught me not evolution itself but a precursor to understanding it: the nonuniqueness of the human species in the continuity of animals. Darwin himself expended great effort to the very same end. Parts of *The Descent of Man* and *The Expression of the Emotions* are devoted to narrowing the gulf between us and our cousins. What Darwin did for his adult Victorian readers, Dr. Dolittle did for at least one small boy in the 1940s. When I later came to read *The Voyage of the Beagle*, I fancied a resemblance between Darwin and Dolittle. Dolittle's top hat and frock coat, and the style of ship that he incompetently sailed and usually wrecked, showed him to be a rough contemporary of Darwin. But that was only the beginning. The love of nature, the gentle solicitude toward all creation, the prodigious knowledge of the science of life, the scribbled descriptions in notebook after notebook of amazing discoveries in exotic foreign parts: Surely Dr. Dolittle and the "Philos" of the *Beagle* might have met in South America or on the floating island of Popsipetel (shades of plate tectonics), and they would have been soul brothers. Dolittle's Pushmi-Pullyu, an antelope with a horned head at both ends, was scarcely more incredible than some of the specimens discovered by the young Darwin. When Dolittle needed to cross a chasm in Africa, swarms of monkeys gripped one another by arms and legs to constitute a living bridge. Darwin would instantly have recognized the scene: The army ants he observed in Brazil do exactly the same thing. Darwin later investigated the remarkable habit among ants of taking slaves, and he, like Dolittle, was ahead of his time in his passionate hatred of slavery among humans. It was the only thing that roused both these normally mild naturalists to hot anger, in Darwin's case leading to a falling-out with Captain Fitz Roy.

One of the most poignant scenes in all children's literature is in *Doctor Dolittle's Post Office:* Zuzanna, a West African woman whose husband has been seized by the wicked slaver, is discovered all alone in a tiny canoe in midocean, exhausted and weeping, bowed over her paddle after having given up her pursuit of the slaving vessel. She at first refuses to speak to the kindly doctor, assuming that any white man must be as evil as Davy Bones. But he coaxes her confidence and then summons up the resourceful fury of the animal kingdom in a successful campaign to overpower the slaver and rescue her husband. What irony that Hugh Lofting is now banned as a racist by sanctimonious public librarians! There is something in the charge. His drawings of Africans are steatopygic caricatures. Prince Bumpo, heir to the kingdom of the Jolliginki and an avid reader of fairy tales, saw himself as Prince Charming but was convinced that his black face would frighten any princess wakened by his kiss. So he persuaded Dr. Dolittle to mix a special preparation to turn his face white. Not good consciousness-raising, by today's lights, and hindsight can find no excuse. But Hugh Lofting's 1920s simply were racist by today's standards, and of course Darwin was, too, like all Victorians. Instead of being smugly censorious, we should look to our own accepted mores. Which of our unnoticed isms will the hindsight of future generations condemn? The obvious candidate is speciesism, and here Hugh Lofting's positive influence far outweighs the peccadillo of racial insensitivity.

Dr. Dolittle resembles Charles Darwin also in his iconoclasm. Both are scientists who continually question accepted wisdom and conventional knowledge, because of their own temperament and also because they've been briefed by their animal informants. The habit of questioning authority is one of the most valuable gifts that a book, or a teacher, can give a young would-be scientist. Don't just accept what everybody tells you—think for yourself. I believe my childhood reading prepared me to love Charles Darwin when my adult reading finally brought him into my life.

My own acquaintance with Darwinism began disgracefully late. I must have been at least sixteen. For many others it is even later, and for most people it is never. Every child in Christendom learns about Adam and Eve and the six days of creation. Some are taught it as literal truth, which is an educational outrage. Others learn it as allegory or myth, which is harmless enough—but so disappointing, so demeaning and shallow, to be taught nothing but allegory, given that since 1859 the stunning truth has been available to us all. There is no reason why evolution should not be taught to seven-year-olds. They would lap it up.

For my secondary education I was fortunate to be sent to Oundle School when the influence of its celebrated former headmaster, F. W. Sanderson (1857–1922), who had made it the foremost scientific school in England, had still not quite died out. ("The Joy of Living Dangerously" is my tribute, in *A Devil's Chaplain,* to Sanderson.) Here is an extract from one of his sermons in the school chapel. When was the last time you heard anything like this in a religious service?

> Mighty men of science and mighty deeds. A Newton who binds the universe together in uniform law; Lagrange, Laplace, Leibnitz with their wondrous mathematical harmonies; Coulomb measuring out electricity. . . . Faraday, Ohm, Ampère, Joule, Maxwell, Hertz, Röntgen; and in another branch of science, Cavendish, Davy, Dalton, Dewar; and in another, Darwin, Mendel, Pasteur, Lister, Sir Ronald Ross. All these and many others, and some whose names have no memorial, form a great host of heroes, an army of soldiers—fit companions of those of whom the poets have sung. . . . There is the great Newton at the head of this list comparing himself to a child playing on the seashore gathering pebbles, whilst he could see with prophetic vision the immense ocean of truth yet unexplored before him. . . .

A typical Sanderson innovation that survived into the Oundle of my time and should have inspired me, but somehow didn't, was the Week in Workshops. We dropped all normal schoolwork in order to spend an entire week each term in the school workshops. These workshops, which had been Sanderson's pride and joy, were the best equipped of any school in the country, and that may have been part of the problem. I learned how to use an advanced lathe but not how to improvise with bits of string and scraps of metal. Sanderson would surely have been chagrined, and I suspect that my father, an improviser of genius, actually was. One term, I managed to break away from the regimentation that had begun, by the time my cohort came along, to threaten Sanderson's ideals. In a corner of the otherwise high-tech metal shop, I apprenticed myself to a wonderful old blacksmith who taught me his ancient craft and how to weld with an acetylene torch. Sanderson would have been delighted, and my father actually was.

My inspiring zoology teacher, Ioan Thomas, had applied for the job at Oundle because he admired Sanderson, and it showed in his teaching. In *A Devil's Chaplain* I recall the lesson in which Mr. Thomas taught us, by unforgettable example, to say "I don't know." Matt Ridley, author of *Nature via Nurture* and other excellent books, has strikingly conveyed this scientific virtue in a recent article in *The Spectator:* "Scientists are not interested in facts. What they like is ignorance. They mine it, eat it, attack it—choose the metaphor you prefer—and in the process they keep discovering more ignorance."

I recently was walking on a pebbly beach with Ridley and his nine-year-old son, Matthew. The pebbles reminded me of Newton's remark, mentioned in Sanderson's sermon, and I fumbled a poor attempt at quotation. Matt and I were struggling unsuccessfully to dredge Newton's words out of our rusty memories when we became aware of a quiet, shy, murmuring coming from below us. We stopped walking to listen, spellbound. It was little Matthew, and he was word-perfect: "I seem to have been only like a boy playing on the seashore, and diverting myself in now and then finding a smoother pebble or a prettier shell than ordinary, whilst the great ocean of truth lay all undiscovered before me." A century ago, F. W. Sanderson might have taken such a boy as par for the course, but in 2003, knowing today's schools, I was dumbfounded. Another secret reader, and in the television age, too?

Ioan Thomas gave me extra tuition in his spare time, and he got me into Oxford, which I think was the tipping point of my life. I originally applied to read biochemistry. Thank goodness they turned me down and said I should read zoology instead. Zoology at Oxford was taught almost as a literary subject—largely as a series of controversial questions on which we were expected to form a judgment and write an essay, after reading the original research literature. Nothing could have suited me better. I was back to books again, and I had the run of one of the great libraries of the world.

Oxford, especially its unique tutorial system, was the making of me. I had a weekly meeting alone with a tutor, who spent an hour discussing the essay I had crafted for him. I was in intellectual heaven. Having spent a whole week in the library, reading not textbooks but the research literature on some detailed narrow topic, then writing my socks off on the same specialized topic, I felt like the world authority on it. What a wonderful feeling! What a privilege for a nineteen-year-old!

My college mentor recognized that my bias in biology was more philosophical than his own and arranged for me to have tutorials for a term with Arthur Cain, the effervescently brilliant young star of the department. Far from aiming his tutorials at examination success, Dr. Cain had me read only books on the history and philosophy of science. It was up to me to work out the connections to zoology. I tried, and I loved the trying. I'm not saying that my philosophic juvenilia were any good—I know now that they weren't—but I have never forgotten the exhilaration of writing them down.

In my penultimate term, I was sent for tutorials to the great Niko Tinbergen, who eleven years later would win the Nobel Prize. Each week my reading assignment was an unpublished doctoral thesis by one of Tinbergen's students. My essay was to be a kind of examiner's report (delicious presumption for an undergraduate) plus my suggestions for follow-up research, my own review of the history of the subject in which the thesis fell, and my own discussion of the philosophical and theoretical issues raised. Never for one moment did it occur to either Tinbergen or me to wonder whether this assignment would be directly useful in answering some exam question—that, again, would have delighted old Sanderson, who despised teaching with an eye toward exam success, although his school achieved plenty of it. My essays led Tinbergen to offer me a place as his graduate student, and then my real Athens experience began.

My entire career since then has been a sort of extended Oxford tutorial, and my books extended essays. I have continued to treat science as something akin to a literary topic, filled with controversies that exercise such verbal skills as I possess; I have been honed, too, by the experience of tutoring young students myself. The libraries of Oxford continue to be havens of refreshment, where I imbibe the literature of science while occasionally making my own modest contribution to it. I cannot talk to animals, like Dr. Dolittle, but I am beginning to understand the makings of their real-life magic.

One Way of Making a Social Scientist

HOWARD GARDNER

HOWARD GARDNER is the John H. and Elisabeth A. Hobbs Professor of Cognition and Education at the Harvard Graduate School of Education. He is also adjunct professor of psychology at Harvard University and adjunct professor of neurology at the Boston University School of Medicine. Among his most recent books are *The Disciplined Mind, Intelligence Reframed,* and *Good Work: When Excellence and Ethics Meet.*

The fields of corporate law and social science research share one curious feature: Almost no child dreams of joining their ranks someday. A career as an athlete, movie star, physician, or even president is much easier to envision. If law is appealing, it is likely to be trial law; if research proves seductive, it is likely to be cutting-edge biology or physics. Yet here I am, at age sixty, a research psychologist for over thirty-five years who has investigated cognitive development in normal and gifted children; cognitive breakdown after brain damage; the nature of intelligence, creativity, and leadership; and the fate of professional ethics in a market-drenched society. If I had not at one point taken an academic turn, I would probably have joined a large law firm and now be contemplating retirement. What insights can I provide about my personal path?

Hitler's mission of ridding Europe of Jews had effects he could not have anticipated. I am part of the third wave of immigrants from Nazi Germany, the first two being those who fled as adults and those who escaped as children. My parents, born in Nuremberg before the First World War, came to the United States in 1938, arriving on November 9, the date of the infamous Kristallnacht in their hometown. They soon moved to the small coal-mining city of Scranton, Pennsylvania, where I was born in 1943 and my sister Marion three years later.

Two events cast large shadows on my childhood. The first was the Holocaust. Like many victims of the Nazi regime, my parents did not talk about it much to my sister and me, or to acquaintances. But their perennial preoccupation was clear from the many stories I heard about individuals lucky enough to escape beforehand, the few relatives who managed somehow to survive the death camps, and the less fortunate who did not. Just recently, I came to realize that my father had led a small brigade that traced the fate of every family member in Europe or elsewhere in the Diaspora. He provided whatever aid he could. Many relatives spent many nights at our small apartment in Scranton, and some even lived there for a while.

The second event was the death of my older brother, Eric. Born in 1935, he arrived in America three years later without knowing any English, emerged as a precocious student, and then died before my mother's eyes in a tragic sledding accident. My mother was pregnant with me at the time. My parents thought they had lost everything; indeed, many years later they told me they would probably have committed suicide if my mother had not been carrying me. Strangely, almost inexplicably to those of us living in the United States in the twenty-first century, my parents did not tell me about Eric. Probably they were just unable to. When I asked about the identity of the child whose photograph was prominently displayed around the house, I was told that he was a child "in the neighborhood." Of course, like all children, I eventually arrived at the truth myself. There is no question that the loss of the gifted Eric and the aspirations my parents transferred to me were important influences on my development, though it would take many years on the couch to unpack those effects.

In the studies my colleagues and I conduct of leaders in the professions, we ask people to reflect on what they were like when they were ten years old or so. If you had shadowed me in Scranton in the early to mid-1950s, what might you have seen? A dark-haired, slightly chubby, bespectacled boy of above average height, who walked and moved somewhat awkwardly. I was a studious sort. I loved to read. I was curious about many things and eagerly peppered older children, teachers, and adults with questions, the more difficult the better. I also liked to write, and by the age of seven or so I was a journalist, publishing my own home and school newspapers. I began to play the piano at that age and was a gifted and serious pianist until adolescence; I might have pursued a musical career—more probably as a composer than a pianist—except that I eventually found practicing onerous. Because of the circumstances of my brother's death, my parents discouraged me from athletic activities. I also suffered from poor eyesight; I had been born with crossed eyes, and I was color-blind, myopic, unable to recognize faces, and incapable of binocular vision. The glasses helped some. I did go to summer camp for seven years and was a dedicated Cub Scout and Boy Scout, winning drill contests and attaining the rank of Eagle Scout by age thirteen. I got a lifetime of camping out of the way before my bar mitzvah in 1956.

With regard to my development as a scientist, what stands out in my early biography is the *absence* of the usual markers—perhaps more surprising in future theoretical physicists or molecular biologists than it is in future social scientists. I did not crave the out-of-doors. I did not go around gathering flowers, studying bugs, or dissecting mice—unless required to do so by the scouting merit badge I was seeking. I neither assembled radios nor took apart cars. I did perfectly well in science and math in school but showed no tendency to pursue those topics on my own; indeed, I was more interested in history, literature, and the arts. And I knew nothing about psychology, though I do remember looking through a psychology textbook when I was a teenager and being intrigued by the discussion of color blindness.

My life changed fundamentally when I entered Harvard College as a freshman in September 1961. I had long been the proverbial big fish in a little pond. Now, for the first time in my life I was surrounded by peers who were at least as able as I was in matters academic and artistic. Daunted at first, I rose to the challenge to become a successful college student. I loved Harvard; it was an Elysian field for the mind. I took many courses and audited more courses than anyone else I knew, running the gamut from Chinese painting to the history of economic thought. I also found my academic interests cohering. Beginning as a history major, I soon discovered that empirical social science questions were more interesting than purely historical ones. I shifted my major to something called social relations, a newly emerging, hybrid field (it never truly emerged) combining psychology, sociology, and anthropology. I was deeply influenced by the charismatic psychoanalyst Erik Erikson, who became my tutor in my junior and senior years, and by other scholars in the social sciences broadly construed—several of whom were of European or Jewish background and representative of the first and second wave of 1930s immigrants.

Whether consciously or not, I took courses that would have been expected of a prelaw or premed student and did well in them. But I had no desire to pursue law or medicine; rather, I wanted to show myself (and no doubt my parents) that I could have pursued those careers had I wanted to. Already I was edging toward a life in the social sciences, probably psychology. At first, under the influence of Erikson, I was attracted to clinical psychology. But once I had encountered Harvard's cognitive psychologist Jerome Bruner and begun to read the powerful writings of Jean Piaget I turned to cognitive developmental psychology. And so, after a year of reading sociology and philosophy at the London

School of Economics, that was the direction I followed in my graduate work at Harvard. At Harvard I came to know the distinguished philosopher Nelson Goodman, who in 1967 established a research group at the Graduate School of Education called Project Zero, which focused on systematic studies of artistic thought and creativity. I was a founding member of the group and have happily remained there ever since, serving for twenty years as its codirector.

What about the "harder" sciences? I was never much attracted by mathematics, physics, or chemistry. I did like biology—it helped to have George Wald, soon to receive a Nobel Prize, as my teacher in college. As a postdoctoral fellow, however, working with neurologist Norman Geschwind, I chose to work in neuropsychology and spent twenty years working in an aphasia clinic. My most important scientific papers are in neuropsychology, where I was one of the first to investigate the linguistic competences of the right hemisphere. I probably could have had a reasonably successful career as a cognitive neuroscientist or perhaps even a developmental neurobiologist, but I eventually left the straight science track and moved to issues of educational reform and social policy.

Could I have excelled as a classical bench scientist? The answer is "Probably not." My talents lie more in the area of synthesis than in the area of innovative experimentation; my research was perfectly respectable but it did not stand out from that of dozens of other researchers. I marvel at investigators like Paul Ekman, who has for decades studied the expression of facial emotions. I could never do that! I sometimes wonder whether, given a different early history, I might have been attracted to the hard sciences rather than the humanities and arts. I always did better in the quantitative than the linguistic sections of standardized tests; however, neither the inclinations of my parents nor the skills of my teachers pushed me in the direction of science, and I did not have strong enough intrinsic motivation to pursue it on my own. What would have happened if I had been raised in the household of, say, biologists George Wald and Ruth Hubbard is anyone's guess!

Given that I ultimately became a researcher and synthesizer in the social sciences, what clues to my career line can I find in my early experiences? Four stand out.

First, I always had a wide and relatively undisciplined curiosity. As a child I liked to read books, newspapers, magazines, even encyclopedias. I particularly cherished biographies. Today, I read more newspapers and periodicals than anyone else I know—and far more than is advisable! This kind of searchlight curiosity is probably more appropriate for social science than the laserlike focus needed by molecular biologists and particle physicists. It also may explain why I have not hesitated to investigate new areas even when cautioned not to do so. I am eager to learn about things that have not yet been described and analyzed and to share my tentative syntheses with others.

Second, my interests have tended to focus more on people and social matters than on the operations of the nonhuman natural world or the world of physical objects. Why this is so is hard to say, though it's also true of most of my family. My guess is that the older generations of my family, though profoundly dedicated to the education of their children, were not themselves highly educated, so their knowledge of science was modest. It was easier and more natural for them to direct their curiosity to the human sphere.

Third, my interest in the humanities has always been characterized by a certain distance. I'm more interested in understanding human beings than in portraying them (as a novelist would) or in helping them (as a clinician or a schoolteacher would). As a member of two marginal groups in Scranton (immigrants and Jews), I was more aware of these "human" issues than the average WASP member of the majority. Yet as someone who had been shielded from painful events—the Holocaust, the death of my brother—I found myself guarded when it came to dealing directly with the pain of human experience. I prefer to investigate them at one step removed. In fact, I have found it almost unbearable to learn about the Holocaust from photographs, films, writing. I was able to watch the movie *Schindler's List* only because of a research project I was involved in, and I recently walked out of *The Pianist* because I found it too painful to observe the inexorable transitions from lives of comfort to lives of discomfort, torture, and eventual annihilation.

Finally, my approach to understanding has typically begun with an effort to define, categorize, create taxonomies. In that sense, as E. O. Wilson once pointed out to me, I approach the human sphere as a naturalist would. Even my earliest papers in psychology and social relations display this classifying bent. My books follow a pattern: the description of an intriguing phenomenon; the development of an approach, which involves the introduction of a taxonomy; and the working out of the approach with reference to a set of examples spawned by the taxonomy. I have followed this procedure in my studies of cognitive science (*The Mind's New Science*), creativity (*Creating Minds*), leadership (*Leading Minds*), professional ethics (*Good Work*), and, most recently, *Changing Minds.* My approach is also primarily descriptive rather than explanatory. One can learn a great deal from careful description (thus speaks the lingering humanist in me) and I am leery of plunging into explanatory models, with their associated bells and whistles. I am also suspicious of the sharp line usually drawn between description and explanation; good descriptions take us quite a bit of the way into explanations. Something else that separates me from most scientists (even most social scientists) is that my preferred mode of expression is the book rather than the article or monograph. I think readily in book form. I like to play out my emerging understanding of phenomena in book form, to lead the reader along the path I followed myself—and to do so in as harmonious and well-constructed a way as possible.

Perhaps because I see myself primarily as a describer and synthesizer instead of a pioneer, I have been surprised to find myself at the center of controversies. I have preferred to work quietly in my study, investigating topics that do not attract the interest of others and avoiding polemics. I was surprised by the strong public and academic reactions (pro and con) to my theory of multiple intelligences—the claim that human beings have eight relatively autonomous intelligences rather than a single one. But I learned that I could engage in debate without losing my bearings. Over a lifetime of reading and reflecting, one reaches strong conclusions. Perhaps this occasional involvement in controversy has allowed me to express some of the latent performer and lawyer traits suppressed by my career decisions nearly forty years ago. Also, I have always been an independent person, unwilling to accept orthodoxy, willing to speak up and defend myself. While I do not relish conflict, I have never run away from it.

By now, some readers will have wondered why, in attempting to make sense of my life line, I have not invoked the theory of multiple intelligences. In fact my ultimate choices do reflect my idiosyncratic configuration of intellectual strengths and weaknesses. Most fundamentally, I am a creature of language and music. I spend my life working with these two symbol systems and expect to do so as long as I am able. I work directly with words, I work while listening to music, and I would like to think that a certain musical sensibility pervades my writing. I am adequate in logical-mathematic pursuits, much less skilled in spatial or bodily kinesthetic endeavors. I have considerable curiosity about the world of other persons, though that curiosity typically involves a certain distance from the more emotional aspects of human life—in this sense I resemble my mentors Jean Piaget and Jerome Bruner rather than my teacher Erik Erikson. As for intrapersonal intelligence, the understanding of oneself—well, that is up to the reader to judge.

If I were facing a career decision today, I think it unlikely that I would elect to go into psychology. Instead, I would search for the current career options that allowed me maximum latitude to pursue my interests in human nature, systematic understanding, and communication with others. And that is precisely the advice I give to young persons who seem dead set on a certain career: "Don't choose the career first; decide what you want to do, and then see which careers are most likely to allow you maximum opportunities and flexibility in the decades ahead."

Let me offer one final perspective. Throughout my writings on education, I have addressed the tension between two important desiderata: the need for years of *discipline*, in order to master any approach to knowledge, and the appeal of *creativity*, the impulse to break out of conventional ways of thinking and discover a new truth about the world. Surely this focus of mine is no accident. From my European—and especially my Germanic—background, I discovered the necessity (and the pleasures) of mastery, be it in the performance of music or the execution of psychological experiments or the drafting of a book. From my life in the United States at a time of notable creativity in many spheres, and because of my somewhat independent and iconoclastic personality, I could never be satisfied just to add one more brick to the current edifice of knowledge. I was prepared to take some chances, in order to possibly break new ground. In this manner, the accidents of individual personality intersected with the conditions of a particular historical era, in the process spawning one social scientist.

someone present, what to do [illegible]. 'Don't choose the career that people think you want to do, and then you [illegible] [illegible] [illegible] – a decade ahead.'

What's the emotional pressure? Throughout my writings on education, I have addressed the tension between two important [illegible]: the need for years of disciplined and [illegible] approach to knowledge [illegible] capable of [illegible] about [illegible] and ways of thinking and discover truth about the world. [illegible] this [illegible] is [illegible] from my European—and especially my German—background, [illegible] the necessity [illegible] the [illegible], be it in the performance of music or the [illegible] of psychological [illegible]. The [illegible] in the United States [illegible] [illegible] individual [illegible] personality could never be satisfied just [illegible] in the [illegible] knowledge [illegible] to take some chances in order [illegible] of [illegible] personally interacted with the conditions [illegible] the process [illegible]

Brains Through the Back Door

JOSEPH LEDOUX

JOSEPH LEDOUX is the Henry and Lucy Moses Professor of Science in the Center for Neural Science, New York University. He is the author of *The Emotional Brain* and, most recently, *Synaptic Self: How Our Brains Become Who We Are.*

When I was a kid, science was far from my mind. Not everyone knows what they want to do at an early age, and I certainly didn't have a clue. I was born and grew up in Eunice, a small town in Louisiana's Cajun country. Most of the prominent grown-ups in my life, including my father, were cowboys. Today I do brain research. I study the way the brain makes memories and emotions. So how did I become a scientist?

My father had been a bull rider in the rodeo in his teens before settling down (sort of) as a butcher. Whenever he could escape from the family meat business, he spent every possible moment working with and riding his horses on our farm on the outskirts of town. He trained them to herd cattle and for weekend horse competitions, and he always had a trail of men of various ages following him around. Some were already cowboys; others were hoping to learn the secrets that would catapult them into cowboyhood. Much to my dad's disappointment, I didn't want to be a cowboy. I didn't have anything against it, but I just wasn't moved by the idea. Being an only child and not into horses, I spent more time with my mother, who liked to visit her sisters or (somewhat better) go fishing, when she wasn't tending to the family business.

One experience I can think of from my childhood that may have pointed me toward a career in brain research occurred on certain Saturday mornings. In addition to trying to get me to be a cowboy, my father did his best to teach me how to be a butcher, so that I could take over the business when the time came. You can imagine how appealing it was for a boy to be asked to roll out of bed before sunrise on Saturday mornings and start slicing meat. The only good thing about it was that I didn't have far to go at that ungodly hour: We lived upstairs above the market. On those mornings, I worked in the back room with three or four men, shady characters who had typically been up most of the night drinking and chasing women—or so they said. I learned a thing or two from their stories, and I even learned a little about butchering.

Other than cutting meat, I had two main jobs in the family business. One was cleaning the pigs' feet. I then took those around the neighborhood on my bike and sold them for a nickel apiece, which supported my baseball card habit. My other job was cleaning the cows' brains. Now we're getting somewhere!

An actual brain is a soft mushy mass, with the consistency of Jell-O. But it's only like this after you remove the tough membranes encasing it, which adhere to its surface and have the texture of coarse stockings. They do an excellent job of protecting the brain, and they are not so easy to strip away. But with patience, the membranes can be peeled from the surface and from all the nooks and crannies, exposing the blob of Jell-O. Then you have to run your fingers into the blob to track down and extract the bullet. In those days, and maybe still today (I have no idea), cows were killed with a single shot to the head. Removing the bullet was a very important task, since customers were not fond of chomping down on lead while enjoying their sweetbreads.

It takes a certain emotional distance to run your fingers through a brain. You have to put aside any idea that the cow's brain was the home of the cow's mind, and just treat it as a piece of meat. It's not as troubling to think of removing a bullet from, say, a pancreas as it is to remove a bullet from a brain. Body organs are not that different from one another in the butcher's sink. But they can have a different impact on the butcher's mind—at least they did for me. I couldn't help trying to imagine what the cow experienced the moment the bullet penetrated its brain. Did its life flash before its eyes? Did it contemplate the bovine afterlife?

These thoughts about cow death on early Saturday morning stemmed less from my being a budding philosopher than from my intense immersion in Catholicism at the time. Most people in southern Louisiana are devout Catholics, and my parents were no exception. I attended a school run by nuns, who fed me deep dishes of theology and convinced me that a life as a priest was the life for me. I bought it for a while, but then something important happened in my own brain: puberty!

My fervor shifted from religion to the two passions of teenage boys: girls and guitars. From that point on, I had one focus: escape from Eunice! I was sure that college was the answer. But by the time I was old enough for college, Louisiana State University had opened a campus in Eunice. The major accomplishment of my young life was convincing my parents that they and I would be better served by the superior education I would receive at the main campus in Baton Rouge. They agreed, as long as I majored in business and promised to come back to Eunice as a banker. I was desperate. I promised.

Although I majored in business, I had no interest in it. This was the 1960s, and Nader's Raiders were hot. Business was evil. Consumers needed protection. So I studied consumer psychology and marketing. I made it to graduation, but not wanting to go to Vietnam I continued in graduate school, getting a master's degree in marketing. It may appear that I was getting farther and farther away from brain research. But though I didn't know it, I was on my way to my life's career.

While studying marketing and consumer behavior, I had begun taking psychology courses, and I ended up with a minor in psychology. One of the courses I took, entitled "The Psychology of Learning and Motivation," was taught by Robert Thompson, a well-known brain scientist who was the resident biological psychologist at LSU. Thompson was trying to achieve what his idol and mentor, Karl Lashley, the father of biological psychology, had failed to accomplish: the localization of memory to specific circuits in the rat brain. I was so taken with Thompson and his research that I volunteered to work in his lab. He took me under his wing and encouraged me to apply to PhD programs in biological psychology. With his help, I was accepted at the State University of New York at Stony Brook, where I received my PhD in psychology.

Guided and inspired by my PhD adviser and friend Michael Gazzaniga, now director of Dartmouth's Center for Cognitive Neuroscience, I took a key step toward my current career, performing my thesis work on patients who had undergone split-brain surgery in order to control their epilepsy. This research—in which one took advantage of the split arrangement in the brains of those people to ask them questions about how one hemisphere felt about emotions, thoughts, behaviors and conscious experiences of the other hemisphere—is what got me interested in how emotions are processed and stored as memories outside of consciousness, the topic I work on today. As a postdoc at Cornell Medical Center in the late Donald Reis's neurobiology laboratory, I acquired the on-the-job training I needed to become a neuroscientist with skills in anatomy, physiology, and brain chemistry. Reis, too, was a great mentor, and I learned much from him.

To those young people who are interested in pursuing a career in science, I don't necessarily recommend the path I took. There's much I've had to learn on the fly. But it has all been an interesting adventure—and fortunately it worked. I guess it's never too late to find yourself. Forty years later, I'm still trying to find myself on the guitar.

The Objects of Our Lives

SHERRY TURKLE

SHERRY TURKLE is Abby Rockefeller Mauzé Professor in the Program in Science, Technology, and Society at MIT and the founder and director of the MIT Initiative on Technology and Self. She is the author of *Psychoanalytic Politics: Jacques Lacan and Freud's French Revolution, The Second Self: Computers and the Human Spirit,* and *Life on the Screen: Identity in the Age of the Internet.*

My mother, Harriet Bonowitz, was tall, voluptuous, and glamorous. People said she looked like Rosalind Russell. By the time I was eight, she was homebound in Brooklyn, with three children and my stepfather, a civil servant who worked as a waiter on weekends so we could afford school clothes and summers at a bungalow colony at Rockaway Beach. My mother's ambition, her heart's desire, was not so much financial as flamboyant—less a house in the suburbs than an evening at the Copacabana. Young as I was, I knew she had decided that her best chance to achieve such dreams was to experience them through me. I was clever, and in my mother's boundless love she assumed I could excel at anything. If I had shown the slightest talent for the stage, she would have become a stage mother, mouthing my lines in the wings. But despite my total lack of musicality or dramatic gifts, for several years she was convinced that I would become a nightclub singer.

We lived in the Brighton Beach section of Brooklyn, and for my tenth birthday I was taken to a local supper club, the Eléganté, on Ocean Parkway. When the chanteuse came onstage, my mother nudged me and urged me to pay close attention. I knew she hoped the performance would inspire me, and I remember feeling immune to her exertions. I was armed with a powerful idea: To be good at a job, you had to love the stuff of the job. I had gotten this idea from a book I owned, a book in my home. I remember its title as *How to Choose the Right Job for You* and its introduction was specific: If you love hammers, wood, and tools, think about being a carpenter. If you love makeup and high heels and sexy dresses and pianos (my mother!), think about being a nightclub singer. If you love paper, notebooks, different-colored binders, and schoolbooks from all over the world (myself, even at ten!), think about being a writer. While my mother was imagining me as a nightclub singer, I was imagining myself as a writer. More than that, the idea that I would do best professionally if I felt a connection to the objects I encountered on my job sparked my imagination. Little by little those objects inspired curiosity about what content, intellectual and emotional, they might carry. Today, as an ethnographer and psychologist, I study the subjective side of our encounters with technology; I am concerned with the human meanings of the objects of our lives. The idea that objects have more than instrumental effects has stayed with me all my life.

That I first encountered this idea from a book that I owned was itself unusual, because there were so few books in my home. There was Carl Sandburg's biography of Abraham Lincoln and then there was my personal library. My library had three elements. First, there was the twenty-four-volume Funk & Wagnalls Encyclopedia that my grandmother Edith Bonowitz had bought for me at $1 a volume—the volumes were released one every two weeks—at her local A&P supermarket. Second, there was my stash of Nancy Drew mystery stories—my half of the collection of books, tied up in neat parcels, which my best friend Helene and I had removed from the incinerator room of our apartment building. This was the way books got passed from family to family in our apartment building; most of the families there did not know one another but had similar backgrounds and incomes and knew that others in the building did not have extra cash to buy books for their children. When it came to reading, parents and children alike used the public library, about a half hour's walk from our building. So for Helene and me, finding a pile of the very precious Nancy Drews was dizzying. There were nineteen of them. Helene had made the discovery and called me to the incinerator room, so she got to keep ten and I got to keep nine.

In order to solve her mysteries, Nancy was usually called upon to decode an object: a brassbound trunk, an old clock, a diary, a pair of twisted candles, a moss-covered mansion. Some of the objects had hidden compartments or secret drawers that would suddenly open. Sometimes these secret places were physical—trapdoors that Nancy's deft fingers would find after days of assiduous searching. But sometimes they were metaphorical—secrets that Nancy's deft intellect would unravel after days of turning the object over and over in her mind. I searched my neighborhood for mysterious objects to decode. There was no lack of candidates. But unlike Nancy, my efforts did not make me a heroine. Our neighbors complained that I was prying. My mother was mortified. I told her I would stop, but I made a promise to myself that I would find mysterious objects to decode outside of Brooklyn. Indeed, I have sometimes wondered if this is how my unconscious interprets my work as an ethnographer (I have studied psychoanalysis in Paris and high technology mostly in Massachusetts and Silicon Valley)—as a way of playing Nancy Drew without breaking that long-ago promise to my mother.

The final element of my childhood library, and in retrospect the most precious, was a pair of guidebooks that had belonged to my mother's sister, my Aunt Mildred, and had been passed on to me. The first was *New Horizons World Guide: Pan American's Travel Facts About 89 Countries*, which Aunt Mildred had received from her travel agent when she booked a trip to Spain, her first trip overseas. My aunt, an unmarried woman, a "working woman," was the first person I knew who had ever traveled beyond the confines of Brooklyn, not to mention the confines of the continental United States. Aunt Mildred was exotic and beloved. I wanted to be a working woman. And I wanted a travel agent.

New Horizons contained information on all the countries of the world, including their capital cities, their average temperatures, their forms of government, their best hotels and restaurants, their main sights. I was able to hold on to the copy my aunt gave me, the 1953 edition, for many years as I changed cities, schools, and jobs, but it finally got caught in a flooded basement and had to be thrown out. I found another copy last summer at a flea market in Wellfleet, Massachusetts. This was a moment not only of pleasure but also of great relief; I felt as though something I had been missing had been returned to me. What is most remarkable, as I read it today, is that *New Horizons* makes every country, from Vietnam to Iran, from Pakistan to Cuba, sound accessible, safe, one call to a travel agent away. In order to set up a vacation both cultural and relaxing in, say, Cambodia, one need only get used to pesky changes in currency and communicate some preferences about one's favored mix of sightseeing and spa elements. *New Horizons* discussed the *aguinaldos* of Puerto Rico, tipping habits for French taxis, and sailing and canoeing in Greece. But its larger message, its message to me, was that access to all of these things was matter-of-fact. This book provided more than information. It gave me permission. At a time when a subway ride to Manhattan was still forbidden to me, the left bank in Paris was taking shape as a future certainty.

The second guidebook was, of course, *How to Choose the Right Job for You,* the book whose good counsel had sustained me at the Elégante as my mother poked me in the ribs. This job guide was published by an employment agency that gave it free to all of its clients, which was how Aunt Mildred came by it. If *New Horizons* gave me wanderlust, *How to Choose the Right Job* gave me a sense of differences among people and a first strategy for defending my sense of difference in my family. Beyond this, it got me curious about people's connections to the objects they love. Our apartment was small. My grandparents shared a dresser. Each of them had space for so few personal objects on its surface. How did they choose these objects?

My mother had a Brownie camera, but in my family we did not take many photographs. There was always a lot of talk about the expense of film and the cost of developing a photograph, which meant that a photograph in our home was not so much an object as a special event. There was, too, the unspoken consensus that each family member had a small allotment of photographs to be "spent" on them. When it was my turn to have a picture, I filled it with my special objects. In a photograph of me at ten, I am at my grandparents' home, seated on my grandfather's chair. I am wearing my grandmother's white gloves and holding two of my most treasured objects: a small figurine that Aunt Mildred had brought back from Mexico and a doll she brought me from France. The photograph contains objects that represent the people who most love and ground me, as well as objects that represent the world beyond Brooklyn that my aunt had access to. France and Mexico were my first destinations abroad, representatives of the world into which I imagined myself at ten, and in the photograph I concretized my fantasies with my object treasures.

If my mother had dreams of me on the stage, triumphing on opening night at Ocean Parkway's Eléganté, my grandparents, Robert and Edith Bonowitz, offered a certain refuge because they had no dreams for me at all, and so I could have my own. As far as I could tell, my grandparents wanted only my love and for me to avoid the evil eye, this latter a constant threat because of envious neighbors. My grandmother tried to ward it off as best she could, by frequently spitting on my forehead.

My mother was divorced when I was two. Until she remarried, when I was five, we lived with my grandparents and Aunt Mildred in my grandparents' one-bedroom apartment. I slept on a cot between my grandparents' twin beds and my mother and aunt shared a foldout sofa in the living room. When my mother and I moved out, my aunt had the sleeper sofa all to herself. After we left, I missed my grandparents and my aunt terribly, and until I was about thirteen I had a "sleepover" at my grandparents' home almost every weekend. My weekend sleepovers had a fairly regular routine. On the television there was Perry Mason, Perry Como, Jackie Gleason, and the Hit Parade. On the kitchen table, my grandfather and I built complex houses of plastic snap-together bricks and other architectural elements. My grandfather worked as the manager of a movie house in Times Square that had a very rough clientele. I knew that my grandfather loved to build these plastic-brick houses because he was good at it and because he could watch me discover how good an architect and designer he was. And more important, since we were working on the kitchen table, my grandfather could watch his wife and daughter see him excel in something that took planning and taste. They knew he was a strong and honest man who could handle the rough trade in Times Square, the vagrants who frequented his theaters and whom he affectionately referred to as "my bums." But every weekend the brick cities we built revealed him as a man of refined imagination. I loved the plastic bricks, a simple, transparent material that not only gave pleasure but encouraged and revealed talent. In later years, when I found computer software that had this transparent quality, it gave me pleasure to watch people use it, and when I found computer software that closed down people's sense of mastery and understanding, I approached it with skepticism. I am aware of my bias: I have sometimes described myself as an instinctive modernist in a postmodern world.

So on a sleepover weekend with grandparents and aunt, I would play several rounds of making buildings with snap-together bricks and sit with the grown-ups through several TV shows before I retired to my special place. Space in my grandparents' apartment was limited, but all of the family photos—including my aunt's and my mother's books, schoolbooks, notebooks, and photographs—were stored in a kitchen closet, set high, just below the ceiling. I could reach this cache only by moving the kitchen table in front of the closet and standing on it. This I had been given permission to do, and this is what I did, from age six to age thirteen or fourteen, over and over, weekend after weekend. I would climb up on the table in the kitchen and take down every book, every box. The rules were that I was allowed to look at anything in the closet but I was always to put it back. The closet seemed to me of infinite dimensions, infinite depth.

I never remember a time when I did not find something new in the closet. Each object I found—every key ring, every postcard—received the care and attention that Nancy Drew gave to her objects. In the closet, every high school notebook with its marginalia—some of it my mother's, some of it my aunt's—signaled a new understanding of who they were and what they might be interested in; every photo of my mother on a date or at a dance became a clue to my possible identity. My biological father had been an absent figure since I was two. My mother had left him. We never spoke about him. It was taboo even to raise the subject. I did not even feel permitted to think about the subject.

My grandmother, grandfather, and aunt would sometimes come into the kitchen to watch me at my investigations. At the time I didn't know what I was looking for. I think they did. I'm looking, without awareness, for the one who is missing. I'm looking for a trace of my father. But they had been there before me and gotten rid of any bits and pieces he might have left—an address book, a business card, a random note. They had cut out my father's image from every photograph. Once, I found a photograph with the body still there and only the face cut out. I never asked whose face it was; I knew. And I knew enough never to mention the photograph, I think for fear it might disappear. It was precious to me. The image had been attacked, but it contained so many missing puzzle pieces. What his hands looked like. That he wore lace-up shoes. That his trousers were tweed.

If there is a sense of vocation to become attentive to the detail of other people's narratives, mine was born in the smell and feel of the memory closet. That is where I found the musty books, photographs, high school class notes that made me feel connected. That is where I determined that I would solve mysteries, that I would use objects as my clues to the heart of the matter. That is where I decided that when the objects could not tell a full story, I would find a person willing to talk to me before a voice was silenced—before someone was forever cut out of the picture.

Intellectual Promiscuity

MARC D. HAUSER

MARC D. HAUSER is a Harvard College Professor, professor of psychology at Harvard University's department of psychology, and codirector of the Mind, Brain, and Behavior Program. He is a recipient of the National Science Foundation's Young Investigator Award; for several years, he has been voted by Harvard students as one of the most popular professors on campus. He is the author of *The Evolution of Communication* and *Wild Minds.*

I can't recall a single early childhood event that led to my current interests. I was far more interested in sports, fiction, music, food, movies, and friends. I did, however, have the great fortune of good genes and a great environment. My phenotype is the outcome of a mother who was a compassionate nursery school teacher and a father who was a world-class physicist and one of the most voracious intellectuals I have ever met. My childhood was a Renaissance feast of opera, film, philosophy, literature, travel, food, and science. I didn't know any of this when I was five or ten or even fifteen. I know it now.

My first transformation happened at the end of my freshman year as a premed student at Bucknell University, in Lewisburg, Pennsylvania. I entered with a keen interest in tennis, soccer, skiing, and squash, and even found myself joining a fraternity. Fresh from a highly cultured upbringing, I was a kid who wanted to explore other parts of life. Fraternities seemed foreign, bizarre. Why not give them a try? But that year I had one horrid and one wonderful experience that together changed my attitude toward things academic. The horrid experience was a class in molecular biology so myopically focused on grades and memorization that I was immediately turned off. At about the same time, I met a wonderful mentor on the squash courts. Doug Candland was a professor of animal behavior, and he was teaching a course in the spring term of my freshman year. I took it. Doug introduced me to *Sociobiology,* the newly published treatise by Edward O. Wilson, who was the leading promoter of a new vision of life that included selfish genes, adaptive sex ratios, optimal foraging, eusociality, infanticide, deception, and cooperation. I couldn't get enough. Suddenly fraternity parties and sports were sidelined for Darwin, Hamilton, Trivers, Williams, Dawkins, and Wilson. Suddenly childhood memories of watching films by Fellini and Tarkovsky; going to concerts of Bach, Mahler, and Berg; reading Joyce and Céline; discussing Wittgenstein and Sartre at the dinner table; and trying to understand the physics that my father carried out at Bell Labs surfaced. I felt ready to integrate my past with my newfound passion. It is a passion that has never flagged.

As I read Wilson's *Sociobiology,* I was both impressed by the theoretical power of this new approach and surprised by the lack of connection with matters more psychological. Wilson spoke of animals acting strategically but said that this was merely a metaphorical use of the term. The same held for Richard Dawkins's selfish-gene metaphor. But what if there was more to it than metaphor? What if animals could plan, reflect, and cheat? Not only did these questions occupy me but they fueled my interest in experimentation. I knew little about experimental design, but I immediately started imagining tests. If females have preferences for certain kinds of males, then we should be able to manipulate male characteristics to test the causal significance of a long tail, a deep voice, a red face, a big body. We could attach a longer tail, play back vocalizations with different frequencies, alter the color of a face, supplement a diet. Thinking about experiments was easy, enjoyable, and soon an obsessive preoccupation.

Doug's course had another influence on me—one I neither anticipated nor appreciated at the time. It tapped the part of my brain I owe to my mother's genetic contribution: a desire to teach. If memory serves, we handed in only two graded assignments for the entire course. The first was an essay. When Doug gave the essays back, mine had no grade and lots of comments—one of which was "See me." I approached Doug and asked why he hadn't given me a grade.

"You can do better," he said. "Rewrite it."

I did. He handed me back another ungraded essay, full of red ink. "Do it again," he said. I was annoyed but obligingly revised. I threw out the original essay and started over, thinking harder about the problem. This time Doug gave me a grade and, with the exasperation that a pen can reveal, scribbled, "Finally!"

The second assignment was a final exam, open-book and open-notes. Everyone arrived nervous. We took out our blue exam books and Doug wrote the following question on the board: "You are having a dinner party. You are inviting some of the leading figures in the field of sociobiology. Who will they be and what will they discuss? Good luck." He turned from the blackboard, smiled, and sat down. The grumbles and expressions of bewilderment erupted. I can recall thinking that this was the best question I had ever been offered. It was playful and challenging. It asked us to use our knowledge to make arguments, not to regurgitate facts. It was a question that a great teacher would ask.

In the summer before my senior year, I decided to do an honors thesis on paternal investment in primates—a topic of some interest, since human males tend to play a relatively large role in raising children whereas most monkeys and apes do not. I traveled to south Florida, to a tourist attraction in Dade County called Monkey Jungle. There I worked on paternal behavior in pigtail macaques and made a bit of pocket money cleaning the cages and feeding the animals. I also read—mostly work in cognitive science, animal communication, and the philosophy of mind and language. Two papers stand out in my memory, not because they had that much impact then but because they did later on: a piece by Peter Marler on the similarities between birdsong and human speech and one by Robert Seyfarth, Dorothy Cheney, and Marler on the alarm calls of vervet monkeys. In addition to my readings, an encounter with a female spider monkey (detailed in the opening pages of my book *Wild Minds*) also initiated an intense curiosity about animal emotion and thought, a curiosity that is still with me.

I returned from Florida to begin my senior year and make plans for graduate school. My grades were anything but stellar; I had spent too much time divided between my independent studies on the one hand and fraternity parties, friends, and sports on the other. But I was admitted to the University of California at Riverside, to an animal behavior program with the primatologist Ray Rhine and the comparative psychologist Lew Petrinovich. Two experiences stood out in that first year, one a fortunate encounter with another graduate student and the other an opportunity to taste the riches of life in academia. At an afternoon orientation, I met Fritz Tsao, a graduate student in physiological psychology. We immediately hit it off, and spent the rest of the afternoon, evening, and early morning talking about Christopher Hogwood's use of original scores and period instruments, the banality of Sartrean existentialism, the excitement of Jerry Fodor's thinking on language and thought, why the European domination in film was at an end, whether an adaptationist's perspective could ever help uncover the nature of the human mind. A shared interest in single malt whiskey helped the conversational flow. Not only had I found a friend but I had found the best preoccupation of all: thinking. And not thinking within the narrow confines that define an academic field but branching out to all domains of knowledge. We were intellectually promiscuous, and proud of it!

This promiscuous adventure carried over into my interactions with Petrinovich, an expert on songbirds, a boxer, jazz musician, cook, and passionate consumer of and critical thinker about the philosophy of science. I took Lew's courses on Karl Popper, Imre

Lakatos, Thomas Kuhn, and their descendants, spent a season in Berkeley doing fieldwork with white-crowned sparrows, and learned to appreciate the intersection between biology and philosophy. My father had once told me that philosophy was a great hobby but a terrible profession. But here was a way to combine the two. Toward the end of my first year at Riverside, I decided that the program wasn't pushing me enough and moved to UCLA to work with Robert Seyfarth and Dorothy Cheney, the authors of the vervet paper I had read in Dade County, Florida.

The first year at UCLA was a blur, as I tried to finish two years of course work in one and prepare for my thesis work in Kenya studying vervet monkey vocal communication and development. As I prepared, I also took a few courses in philosophy, and in one of these I met a first-year philosophy graduate student named Colin Allen. On the first day of the course, Colin leaned over and asked if I was a student in philosophy. I said I was in animal behavior, and this puzzled him. I explained that I was interested in the nature of mind, the relationship between animal communication and language, and whether monkeys and apes might have some of the same kinds of cognitive skills as we do. He politely told me I was wasting my time. We battled for a term. I won. He did his thesis on the cognitive abilities of animals, in opposition to the theories of John Searle, Fodor, and Donald Davidson. Ultimately we wrote a paper together on the content of animal concepts, especially of life and death, which was published in a philosophy journal. I proudly told my father that philosophy could be more than a hobby.

I left for Kenya curious, full of ideas, and excited to see Africa after all the years I had spent dreaming about it. Robert and Dorothy picked me up in Nairobi and popped me into the back of a jeep. As we headed out of the airport, we slowed to let a giraffe cross the road. I wasn't in Kansas anymore. But then we stopped off at a French patisserie, a butcher shop serving pâté, and a polo match, where I sipped a Pimm's. I wasn't in Kansas; I was in the midst of the British colonial empire. Robert and Dorothy not only knew how to do first-class research, they knew how to live like royalty. After picking up supplies, we drove to Amboseli National Park, near the Tanzanian border. Our campsite was run by Cynthia Moss, an internationally recognized expert on elephant behavior. Cynthia had spent close to thirty years studying elephants and had developed the most beautiful campsite I had ever seen. Our tents were set amid acacia trees and tall grasses. In the background was Mount Kilimanjaro, and we could hear the soft footfalls of elephants and the passage of gazelles, zebras, an occasional lion. For the next two years, this would be my home and my intellectual playground. Before going back to the States, Robert and Dorothy spent two weeks familiarizing me with the vervets in their study groups and arguing with me about sociobiology, philosophy, linguistics, and evolutionary theory. It was incredibly exciting and intimidating. Every time I made an argument, I felt as though Robert and Dorothy were six steps ahead. But I knew I would learn from them. And I did.

During my two years in Amboseli, I became a scientist. I learned how to collect data, how to think critically about it, how to design experiments, and how to write up what I was discovering. I read all of Darwin's original writings and annotated them profusely. I read philosophy, linguistics, behavioral ecology, and cognitive science. I read the *Guardian Weekly* and listened to the BBC to keep in touch with the world. I also learned a great deal about how I wanted to configure my life. I knew I needed more in my life than science. I needed to feed my interests in music, fiction, cooking, and sports. I woke up every morning at 6:00 A.M. to see the sun rise over Kilimanjaro and drink Kenyan tea. I listened to Bach, Coltrane, and the Talking Heads; I read George Eliot, Jane Austen, Jack Kerouac, and John Irving. I cooked extravagant meals with my Kenyan friend and camp manager, Peter. And I played on the local soccer team, an experience that not only helped my Swahili but kept me in great shape and introduced me to Kenyans and Kenyan culture in a way I would never have appreciated had I remained cloistered in my fieldwork with the vervets.

I went back to UCLA with a mountain of data, three papers in press at good journals, and the approaching date of my thesis defense. Although returning to Los Angeles produced an extraordinary culture shock, I eventually eased into seminars, analyzed my findings, and received a PhD. I also landed the first of two postdocs with Richard Wrangham and a year in Uganda studying chimpanzees.

My year in Uganda played an important role in refining my interests and establishing the kind of scientist I would become. Richard was a good empirical scientist, but he was a great theoretical synthetic scientist. His model of primate social organization transformed the field, and his thoughts on the evolution of aggression were beginning to have a deep impact on a broad range of studies in human nature. Working with Richard pushed me to think about science and scientists—whether there were different types or styles, something akin to natural classifications. I conceived of four kinds: theoreticians, synthesizers, empiricists, and popularizers. Some, such as E. O. Wilson, manage to embody all four. This is rare. Most of us wear one hat and occasionally try on the others. I realized early that my strength was in thinking about theories and coming up with novel ways of testing them. I was an empiricist at heart. But I also enjoyed synthesis—seeing the connection between disciplines that rarely communicate. I would leave the heavy theoretical lifting to others. Working with Richard brought forth another realization: I had little patience for observational fieldwork that only slowly yielded patterns. I wanted to run experiments. I had the patience of a jackrabbit, not a lumbering orangutan. Chimpanzee fieldwork was not for me.

In 1988, I moved to a field station in Millbrook, New York, a satellite of Rockefeller University, for a postdoc with Peter Marler, the grandfather of the field of animal communication. Peter imparted a high gloss to my formal education. He pushed me intellectually and shaped my writing. He taught me how to give lectures and how to teach. He told me to slow down when I needed to be told. And he did it all with the grace and style of a master. My research moved from chimpanzees and Uganda to rhesus monkeys and Puerto Rico. My theoretical interests changed as well, as I began to combine questions of adaptive design with issues concerning the brain and mind. I collaborated with speech scientists at the Haskins Laboratory in New Haven, Connecticut, and developmental psychologists working on language acquisition. I spent five of the best academic years of my life with Peter as a postdoc. He treated me like a colleague and friend. I was ready to graduate, to be launched into full independence.

I applied for jobs and in the end had two competing offers from two diametrically opposed environments. One was the National Institutes of Health, which invited me to help build a new unit of neuroethology, where I would have the freedom that a research institute offers, with no teaching responsibilities and virtually limitless research funds. The other was Harvard University and the department of biological anthropology. I was born in Cambridge while my father was a graduate student in Harvard's physics department; taking a job there would be like coming home. But I was nervous. Harvard was notorious for its consumption and regurgitation of young faculty, and I wasn't keen on becoming its next victim. Fortunately, conversations with Irven De Vore, the grand old man of field primatology and hunter-gatherer studies, and Jerry Kagan, the grand old man of studies of infant temperament and the development of moral behavior, convinced me that I had no choice: I would die at the NIH, they said, stifled by the lack of diversity and the absence not just of students but of the creativity that comes from teaching and struggling to obtain funding.

The rest is history. They were right. I went to Harvard and I have never had a day of regret. Although I started my path to tenure in the biological anthropology department, I ended up with tenure in the psychology department. When I get a call nowadays from a reporter who asks, "How shall we classify you—biologist, psychologist, neuroscientist, evolutionary psychologist . . . ?" I'm always stumped. Sometimes I say "biologist," because I believe that evolutionary principles inform our understanding of the mind. Sometimes I say "psychologist," because I'm interested in the mind, language, concepts, and morality. Sometimes I say "neuroscientist," because I'm interested in how the brain processes information and then spits it out, allowing human thought to go where no other animal has gone before. In the end, I'm intellectually promiscuous—and I'm still proud of it.

Tom Swift Jr. and the Power of Ideas

RAY KURZWEIL

RAY KURZWEIL is an inventor, entrepreneur, and author. He was the principal developer of (among a host of other inventions) the first omni-font optical character recognition software, the first print-to-speech reading machine for the blind, the first CCD flatbed scanner, the first text-to-speech synthesizer, the first music synthesizer capable of re-creating the grand piano and other orchestral instruments, and the first commercially marketed large-vocabulary speech recognition system. He received the National Medal of Technology from President Clinton in 1999. He is the author of *The Age of Intelligent Machines* and the national best-seller *The Age of Spiritual Machines: When Computers Exceed Human Intelligence.*

My grandfather came back from a return visit to Europe in the mid-1950s with two key remembrances. One was the gracious treatment by the Austrians and the Germans, the same people who had forced him to flee Vienna in 1938. The other was a rare opportunity he had been given to handle with his own hands some original manuscripts by Leonardo da Vinci. Both recollections influenced me, but the latter is one I've returned to many times. He described the experience with reverence, as if he had touched the work of God himself. This, then, was the religion I was raised in: veneration for human creativity and the power of ideas.

At the age of five, I decided I would become a scientist. (Steven Pinker expresses skepticism in these pages about such early recollections, but I have witnesses!) Although I used the word "scientist," what I meant by it at the time was "inventor." A project that occupied me for several months was building a rocket ship from the parts in my Erector set, along with other materials I had gathered. I can still recall the feeling of transcendence that envisioning my idea gave me. The invention had more reality for me than the uncooperative pieces I was struggling with. My mission of making the idea real for everyone else had a compelling force—so compelling that I spent as much time as I could working on the project.

I remember this urgent feeling because it is an emotion I know very well as an adult. An idea takes shape and just as quickly takes over my life. I still don't really understand this process. I can build a philosophy around it—having to do with the importance of knowledge and understanding—but any such philosophizing is an after-the-fact rationalization. The compelling and insistent reality in my life is the imperative of turning an idea that is already very real for me into a reality for others.

The rocket ship never got off the ground, so I turned to more practical vehicles like go-karts and boats. My go-kart was smashed by a neighborhood bully, but a boat I constructed fared better. I was able to manipulate its path through the water using control strings from the shore. When other kids were wondering aloud what they wanted to be, I remember having the conceit that I knew what I was going to be. By the age of eight, my inventions became more intricate and more practical, such as a robotic theater with linkages to move scenery and characters in and out of view, a mechanical baseball game, and a magic box involving mirrors, in which things seemed to disappear.

Somewhere along the line, I had the notion that inventions could change the world. This optimism was influenced by my discovery, at the age of eight, of the Tom Swift Jr. series. The concept in each of the thirty-three books in the series (only nine when I started to read them in 1956) was always the same: Tom would get himself into a terrible predicament. The fate of Tom and his friends, and often of the rest of the human race, hung in the balance. Tom would retreat to his basement lab and think about how to solve the problem. The moral of these tales was simple: The power of the right idea will always overcome a seemingly overwhelming challenge. I continue to value my collection of Tom Swift books—though they have been collected in recent years on eBay and aren't the originals I read as a child. To this day, I continue to be convinced of that fundamental philosophy: No matter what quandaries we face—business problems, health issues, relationship difficulties—there is an idea that can enable us to prevail. Furthermore, we can find that idea. And when we find it, we need to implement it.

My parents were artists. My father was a concert pianist and orchestra conductor. His flight from Vienna was facilitated by a wealthy American patron of the arts who helped talented musicians from Europe escape Hitler and enter America. My mother was (and is today) a talented visual artist. I am often asked how I came to commit so early to science, coming from such an artistic household. But art and science were not so different in the milieu of my family. Both had to do with creating knowledge and creating a valuable pattern from simple materials.

My mother's mother was one of the first women in Europe to get her PhD in chemistry. She lectured in Europe and ran a school for young women (the Stern Schüler) that her mother had started in the nineteenth century. My mother's father was a physician and a colleague of Sigmund Freud (whose grandson Walter once proposed to my mother). My mother's sister is a psychologist who has recently written books on the Jews uprooted by the Holocaust. My father's father was an engineer, and my father's brother was a gifted inventor, producing elaborate machines to automate European factories. George Parker, a cousin of my mother's, was a talented electrical engineer for Bell Labs; his brother Frank was a brilliant intellectual and New York attorney. So it was not easy for a small child to get a word in edgewise at my family get-togethers. The intense and animated discussions were invariably about new ideas, usually those of intellectuals I had never heard of. The way for me to get attention was to have an idea. And since it was challenging to break into the conversation, it helped if the idea had a material form. There was great respect for learning and accomplishment in my family, so any instantiation of knowledge got their attention.

Aside from building things, I had a passion for information. At the age of six or seven, I used to cut out the "Ask Andy" science question column from the *Long Island* (NY) *Tribune.* I still have my scrapbook, filled with now-yellowed clippings in which Andy would explain why an airplane could fly or why the eye of a hurricane was still. One of my favorite possessions was a deck of cards about the states of the United States and another about countries of the world, listing their key economic and demographic data. When I was ten, I would send letters to state governments asking them for similar information and receive packages with extensive charts and tables. I hoarded all these data as if they were jewels. I had a sense that information was valuable—that precious patterns could be found in all that data.

At age twelve I became fascinated with electrical switches and lights. I built my own circular switches—dials that could connect an input to one of ten possible outputs—and created a calculating system that could perform a variety of computations using tiny lightbulbs for output. There was something missing, however, in that I was unable to make this system really think on its own. It was then that my Uncle George gave me some surplus electrical relays from Bell Labs and explained how they worked. A relay could be wired in such a way as to remember that it had been signaled. Using another pattern of wiring, one relay could control the actions of another.

This encounter with the electrical relay was a true epiphany for me. It became immediately clear to me how one could combine relays to create extensive systems with memory and logic. By hooking together a sufficient number of relays, you could create systems that analyzed problems on their own. If they were hooked up in appropriate ways, they would cause chain reactions of relay signaling.

I began to hang around the surplus electronics stores on Canal Street in Manhattan (they're still there) and gathered parts to build my own computational devices. During classes back home in Queens, I would prop up the textbook prominently on my desk, but underneath I was sketching out increasingly elaborate relay circuits. I built relay-based systems that could solve logical problems, such as directing an electromechanical mouse I had built to find its way through a maze. The time it took for the relay chain reactions to settle down seemed eerily similar to thinking.

Shortly after being introduced to the electrical relay, I discovered the idea of computation. Here the power of memory and logic was organized in a fashion that could easily be controlled through software programs. I was struck by the similarity between my own thinking and what could be modeled on a computer.

In high school, I had a summer job at Flower and Fifth Avenue Hospital in Spanish Harlem, as part of a team of about a dozen people who were computing statistical analyses for research on the Head Start program. We used electromechanical calculators and paper spreadsheets to methodically perform the sequence of steps required to compute such statistical tests as a four-way analysis of variance. I discovered that there was an actual computer—an early minicomputer, the IBM 1620—in the hospital, and I managed to get after-hours access to it. As an experiment, I wrote a program to perform the same computations that our team was working on. I surprised my supervisor by returning a completed analysis, which had been expected to take several weeks to perform, in a couple of hours. My job quickly changed to computer programmer, and I began programming the rest of the statistical tests. I'm not sure what happened to the rest of my calculator team.

After a few weeks of this, I got tired of programming the orderly patterns of statistical tests and used the hospital's computer after-hours to find patterns in musical pieces and create music compositions based on those patterns. My father took a keen interest in my automated composition project, and we talked about another potential role of computers, as the actual creators of musical sounds. He had an interest in Robert Moog's analog synthesizers of the 1960s and agreed with me that a digital computer could do a better job of creating musical sounds than an analog synthesizer. He predicted that someday I would apply my interest in computers to this concept, a task I would return to two decades later. Both parents not only supported my ideas but also the notion of the ascendancy and priority of realizing them. Around this time, my father had a major heart attack, but despite the financial and emotional toll exacted by his illness, he and my mother continued to provide the resources for my increasingly elaborate and expensive projects.

The experiment in music composition was my first attempt to apply computers to pattern recognition. I became fascinated with the idea of modeling the chaotic mental process of recognizing patterns. Although computers had a reputation for being good at orderly sequences of logic, it was clear to me that they could emulate the more unpredictable process involved in distinguishing designs and relationships in the real world. Looking at the often illogical way the people around me thought about things, I was fairly sure that this irrational and self-organizing process lay at the heart of human thinking.

I began to read about artificial intelligence, pattern recognition, and neural nets, which were simple models of how biological neurons were thought to work. I wrote a letter about some ideas I had on these subjects to Marvin Minsky at MIT and Frank Rosenblatt at Cornell. Minsky was one of the founders of the field of artificial intelligence, cofounder of the MIT Artificial Intelligence Laboratory, and an early pioneer of neural nets. Frank Rosenblatt had invented the Perceptron, a classic form of neural net, in 1957 and was regarded as a leading advocate of self-organizing systems at the time. Both wrote back enthusiastically and invited me to visit them, which I did. Professor Minsky continued to be my mentor when I subsequently went to MIT in 1965, and we remain close colleagues and friends today.

Although it was a gradual awakening, I had the idea in high school that a computer could emulate human thinking and could also simulate visual and auditory phenomena and environments. It was already apparent to me that computer technology was evolving rapidly, and that we would ultimately have the tools to re-create our own thinking and to simulate any other complex process found in the natural world.

The power of an idea—this is itself an idea—continues to animate my thinking and my work. I find myself a servant to ideas, but they have rewarded me in return. For example, I was diagnosed with type 2 diabetes more than twenty years ago—a disease that undoubtedly contributed to my father's illness and premature death in 1970. But the right ideas—ideas in direct contradiction (at the time) to the mainstream advice of the medical profession—have effectively overcome this condition. My father made lifestyle changes readily; when it became apparent to him early in his life that smoking was a health danger, he simply dropped the habit overnight. If he had had the knowledge then that I have now, he might be alive today.

Over the last quarter century, I have come to appreciate an important meta-idea: that the power of ideas to transform the world is itself accelerating. Although people readily agree with this observation when simply stated, very few people truly appreciate its profound implications. Within the lifetimes of the majority of people now living, we will have the opportunity to apply ideas to conquer age-old problems. I am convinced, for example, that we will eliminate human aging within the next few decades while at the same time vastly extending our physical and mental abilities. This is the paramount idea to which I find myself, in recent decades, both master and servant.

A Day in the Life of a Child

JANNA LEVIN

JANNA LEVIN is a professor of physics and astronomy at Barnard College of Columbia University and recently held a fellowship from NESTA (National Endowment for Science, Technology, and Arts) at the University of Oxford. She is the author of *How the Universe Got Its Spots: Diary of a Finite Time in a Finite Space.*

My dad had these enormous medical reference books. After he left for work in his green Volvo to drive through rush-hour traffic into Chicago, I'd spend time in his office, in his brown leather armchair with the matching ottoman and the wall of enormous books. I could barely lift them; they filled my whole arms, fingers spread wide on either side of the solid clothbound covers—maroon, dark blue, beige. Ugly colors, which impressed me. Such an insistently drab exterior seemed to signify something, to lend gravity to the contents. That was as much as I could glean from those books. Inside were hieroglyphs. I tried to read the entries—*Pancreatic Adenocarcinoma, Cerebellar Vermis Hypoplasia, Myeloid Myelodysplastic Syndromes*—but never understood a significant word. I was jealous that someone else could access the ideas and knowledge trapped in the Latin inscriptions. Now I think they should have been better written—and more attractive.

I knew that those drab, heavy books were somehow connected to a real place—a *hospital.* I went there with him once and was witness to open-heart surgery. I stood close enough to cause damage when they leveraged the ribs to open the chest cavity and expose the heart. My dad watched alongside me, wondering if maybe this wasn't such a good idea. The surgeon glanced down at me a couple of times—I was just a pair of enormous, shocked eyes between green paper hat and green paper mask—and asked whether I might faint. But I wasn't even close to fainting. I was able to detach myself from any instinctive empathy that makes us imagine that this could hurt. It was a toughness that faded with age, along with any impulse to become a doctor. But that was an extraordinary day; on an ordinary day, I would sit in my dad's armchair and pretend to read his books.

After reciting encyclopedic entries from the Physicians' Desk Reference, I would rummage through the music selection in my dad's study, which was also home to the family record player and stereo equipment. We had a collection of 8-track tapes, huge plastic boxes. They were real machine parts, delivering a satisfying kick when properly locked into the tape player. I'd have to put some weight into pushing the 8-tracks into place. I could sit there for hours listening to Willie Nelson and Rod Stewart, or if I got my hands on my sisters' collections, maybe the Beatles or Queen. But I never became a musician. Or a sound engineer. Or a medical doctor.

My dad's study was a small brown room off a bigger, browner room—the family room. Brown and yellow. There were wood floors and wood paneling. It was the 1970s. And black-and-yellow plaid couches—two plaid couches facing each other. By late morning I'd be sitting between the couches, on the brown rug, as if in a sandwich with giant plaid bread—a Janna sandwich. I'd sit there and watch TV. Hours and hours of TV. I watched reruns of *Star Trek*. I had a crush on Captain Kirk. He would stagger around those chintzy sets, leading with his chest, throwing his chest left and right, his feet following that chest choppily, all barrel-chested, arms bent, fists clenched away from that yellow-clad chest. I loved that show. It was so optimistic. We had flown to the moon! To the moon, can you imagine? The pride, the humility! Suddenly the colossal expanse of the solar system, the intimidating, humongous cosmos, assumed a new significance, and the old significance sped through the social order, from religious mythology past astrology, through science, to real, clunking exploration. We were the new Spaniards pushing off the new coast. We looked again at the plane of planetary orbits and saw our cosmic backyard—a reachable, traversable, knowable backyard. If the solar system was manageable, maybe so was the galaxy, or the whole universe. Our eyes glinted with megalomaniacal madness. All of it could be ours to explore and who knew what was out there? It was science fiction, but it was reality, too.

I watched TV all the day long if it wasn't a school day. *Dr Who. The Land of the Lost. 2001: A Space Odyssey. Cosmos.* We were a new generation. I sat there, a child from a childish country, wide-eyed and absorbent. Knees crossed on the brown floor in the brown room, face illuminated by that artificial flickering light, I soaked the images in through my pupils and pores.

But I also watched old Laurel and Hardy and Abbott and Costello movies and *Bewitched* and *I Dream of Jeannie* and lots of sitcoms about housewives. Jeannie wanted to marry an astronaut she called Master. I fantasized about becoming an astronaut. I grew up in love with space and the cosmos, but not with a man called Master.

In the late afternoons I'd go with my mom to shop for groceries. I was like luggage and would be placed along with her handbag on the orange plastic seat on the front of the shopping cart, legs fitting between metal shelf and handrail. The grocery store was huge and suburban. I don't remember the name of the chain. Maybe it was a Piggly Wiggly or a Safeway. Racks of fluorescent tubes aligned with the aisles, so that the place was bleached with photons. Sometimes, I'd walk alongside the shopping cart paying special attention to the nearness of the slick, shiny linoleum floors—a mysterious, modern material. We would walk slowly and conscientiously down the corridors of food, of cardboard boxes and seductive packaging. It was a waltz. The Piggly Wiggly waltz. As the Muzak urged us on, we'd pack the groceries into paper bags—distinctly American, big, stiff, brown paper bags. This was pop art, consumerism, pure Warhol waiting for me to discover him.

But I never did learn any practical skills there. I can come back from a hangar-sized grocery store with oddly mismatched, overpriced sundries that refuse to merge into a meal. I learned instead about Warhol and his America. A decade later, in New York, I would stand beneath giant painted cans of Campbell's soup feeling happy with the familiarity of the image. Thinking I never much cared for the chicken variety, thinking about massive suburban grocery stores and brown paper grocery bags, thinking about red and the excellence of the label and the stupidity of our consumerism. Thinking I could go for some soup.

Then there was Carl Sagan. After Dad came home, the family would sit around the kitchen table and talk—but not about the hospital or the kids in intensive care. We'd talk about things at random. If the conversation needed flint, there was always Carl Sagan. My whole family used to laugh to the point of derangement at my impersonation of him in *Cosmos,* ogling the animated sky with openmouthed wonder and reveling in the magnificence of the sound of the phrase "billions and billions" and the awesome significance the words denoted. My imitation was sheer flattery; we all thought he was great.

I read his book *The Dragons of Eden* and wrote essays for school on evolution—essays with elaborately decorated covers in careful, childish font and light pencil portraits of *Australopithecus.* A child's scrawl and careful prose, one clean, basic sentence steadily following another—which, come to think of it now, is a better way to write than the way I wrote as a degree-confirmed scientist from places like MIT. I had to unlearn those tongue-tangling polysyllabicisms. My early essays on evolution and astronomy became mementos of a little girl I couldn't remember, mementos I hoped would explain how I got here from there. They never did. I carted those reminders around with me in unopened boxes until my postdoc at the Center for Particle Astrophysics in Berkeley, when, in impulsive frustration, packing for yet another academic move, I emptied them into a recycling bin on Haight Street.

In the evenings, my mother would read books, so many books. She'd settle on the couch and stretch her toes away from the back of the novel. If she finished a book, she managed to find a place for it on already full bookshelves. One day, nosing around, I saw that they were stacked three or four deep on the shelves: Toni Morrison, Philip Roth, Joyce Carol Oates. It would be many years before I would succumb to such capacious reading—an avarice on the verge of addiction, including compulsive spending in bookshops and hoarding of books to ensure that the supply never diminished. Books in drawers, under the bed, in boxes under the stairs. Books high and low and for sheer pleasure. Lust almost. Many, many years later, I would write a book. The intimacy of those books would migrate from memory to influence to experience, until they were digested and stored in my fibers, an integrated part of who I am.

At night I would stay up late, drawing pictures in my room, sprawled out on carpeting so ugly my mother wept after it was installed. It was an irregular pattern of shag and short fibers in candy-colored pink, blue, red, and white. Deep into the hours when it was dark and quiet, I'd lie in bed and watch the clock, a bulbous white plastic brick with numbers painted on black plastic cards. When the minutes changed, a card with the next number would flip down. I watched the time change and played combinatorial games with the numbers, isolating primes, finding divisible integers, cataloging rational combinations. Nothing too impressive. Nothing that would match the stuff of legend, like the story of the great mathematician Carl Friedrich Gauss, who as a schoolboy was given the assignment, along with his classmates, of adding all the numbers from 1 to 100 and produced the answer within a few minutes by noticing that the sum was the same as 101 times 50, while everyone else was plodding along with 1 + 2 + 3 + 4. . . .

I'd sit up at the foot of my bed to look out the window onto the backyard. I'd listen to the neighborhood. Far off, I would hear cars or trucks moving along, and there were insects worth listening to, all of it providing a sound track to my late-night solitude. I'd stare at the patch of sky wedged between the trees arching over the neighbor's manicured lawn. I'd wonder how far I was seeing, how deep into space.

And day after day things like this would happen, and at the end of thousands of those days I would be conferred with degrees and jobs and titles and I'd be a scientist. These are the threads I remember, the things I think I know. How is it that these experiences colluded to make me a scientist and not a musician or engineer or doctor or housewife? I have no idea.

But I know I would swell with a feeling like ecstasy at the thought of our pretty, blue planet spinning tamely in a sea of blackness in a cosmos magnificent and huge. There was nothing I could think of that was more important than this, as I looked out that white wood-framed window—a white wood-framed piece of the universe, as important, as momentous, as insignificant as any other. I wanted to see farther. I wanted to fly out the window through the trees, into the thick color of the sky, and merge with whatever I found there. Night after night. Steaming the windowpane, my face was so close. I wanted to see more, to know more, to *be* more.

When the frustrations of science wear me down, when I can't bear to write another grant application, do another detailed calculation, listen to another seminar, read another article whose title alone is impenetrable, I sometimes wish I had chosen some other direction. I've looked back and wondered how I got here from there. But then after time it always comes around again: I'm still that kid sitting alone in the middle of the night, thrilled just to look through the window at my piece of the universe, wondering what else is out there.

Toward the Worm

RODNEY BROOKS

RODNEY BROOKS is the director of the MIT Computer Science and Artificial Intelligence Laboratory and Fujitsu Professor of Computer Science at MIT. He is also the chairman and chief technical officer of iRobot, a robotics company. He is the author of *Flesh and Machines* and *Cambrian Intelligence: The Early History of the New AI.*

There have been two major turning points in my scientific life. The first was when my parents built a carport on the side of our house.

I grew up in the comfortable, remote city of Adelaide, Australia. Its main role was to support agriculture for a couple of hundred miles around, in an otherwise desert state used by the British for aboveground nuclear tests at just about the time I was conceived and born. Five hundred miles to the east was the more cosmopolitan Melbourne, and two thousand miles to the west was the even more remote Perth. There was not much in between—to the west, not even paved roads.

My father and mother had ninth- and tenth-grade educations, quite respectable for their time; as far as I can recall, none of their friends had finished high school either. My father was a telephone technician, and my mother had been a hairdresser before my older brother arrived. Within that milieu, by age four, I was known as the Professor. I had an uncanny ability to manipulate numbers in my head—to instantly tell you, for example, that 5,347 pennies were 22 pounds, 5 shillings, and 7 pence. I had an obsession with the regularity of arithmetic—with the way numbers made patterns that could be predicted and could be executed within my head through the application of procedures I would devise. I had lots of computational ability, but not much to apply it to.

When the carport was built, it liberated the metal shed at the bottom of the garden from its role as car domicile. Up until then, my brother and I had had to share our father's workbench at the far end of the shed and clean up our woodwork experiments at the end of the day. Now there was room for two more benches, one for Chris on the right and one for me on the left. Chris was interested in chemistry and making noxious gases, and in the interest of complementarity and through some process of family negotiation it was agreed, when I was age seven or so, that my interest would be electricity. My new bench, an old wooden kitchen table, would be the place where I would build electrical experiments.

Once I understood how to make a circuit with a battery and a flashlight bulb, I was drawn to making little logic circuits, with push switches fashioned out of nails and strips cut from tin cans. I could build a circuit that would light a bulb only if switch A *and* switch B were depressed, and another that would light a bulb if switch A *or* switch B was depressed. This was the key to building a mechanism that could do the same sorts of procedures I had been doing in my head. Here was something I could be passionate about: the mechanization of things that humans used their brains for. By age eight, my life's work was determined: I would make machines do things that only people could do by thinking—and I would make those machines do the things that I was very good at doing myself.

I was given an American book, *Giant Electronic Brains,* that described the binary number system and how computers could outperform an abacus expert doing arithmetic. I had never even seen a mechanical calculator, let alone a real computer, but the idea of mechanizing arithmetic was wildly alluring. I soon determined that the complexity of circuits needed to do any serious arithmetic was way beyond my budget; my allowance was only six pence per week—enough to buy two flashlight bulbs at the local Woolworth's. So instead, in order to illustrate machine intelligence, I concentrated on building machines that could play games. By the age of ten or so, I had very little doubt that we human beings were machines in the way we thought, and that emulating human intelligence with a machine was just a matter of circuit complexity. After all, I had read about neurons and the electrical properties of propagating nerve pulses; brains were made of the same sorts of components I was fashioning.

The machine from that era that I most enjoyed building was one that could play tic-tac-toe and not be beaten. At age twelve, I had discovered a source of surplus manual telephone switchboards. They contained three-position switches, which simultaneously made as many as four sets of contacts as the switch level was moved up or down from the neutral position. I used nine of these switches, one for each square on the 3 x 3 board, and had their position encode the contents of the square; neutral for empty, up for an X, and down for a 0. A lightbulb would light in the square where the "computer" (as I called it) wanted to play its first X; the human player would record it as played by moving the switch up, and would then put a 0 in another square by pushing its switch down. In order to design the circuit for this tic-tac-toe machine, I had to manually work out the whole game tree—the set of all possible games that could be played—so that the circuit would pick the right move for the unfolding game. If the machine were given a game position that did not follow from any of the preselected game scripts encoded in its circuitry, it would not know what to do—even if a single move would bring about a win. I knew this was different from the way people played the game, but at the time that was not important to me. I was driven by engineering factors—a mixture of cost considerations and the limitations on my ability to build the more complex transistorized circuits that would have been required to emulate humanlike play. At that stage in my life, although I had developed everything completely independently, I had arrived at a philosophical position remarkably similar to that held by mainstream artificial intelligence (I had not yet heard of that term) researchers in Europe and the United States: namely, that it didn't matter whether a computer used the same techniques as a human to produce intelligent behavior; all that was important was that the surface-level description of that behavior was the same. My computer could play coherent games of tic-tac-toe. I did not care that it could not do equally simple tasks outside the context of the game—or that it had neither the concept of "three in a row" nor even of a two-dimensional board. Now, looking back, I realize that mainstream AI researchers arrived at that philosophical position driven by the same ultimate considerations that drove me to it. And many of them were also seduced by the allure of having their machines do things that they themselves were better at than other people.

My passion for building intelligent machines continued through my teenage years and after a while I mastered transistorized circuits. By the time I entered Flinders University, in Adelaide, I was buying 7400 series integrated circuits and building ever more complex nonarithmetical computers. Finally my mechanical skills got to the point where I was able to fabricate my first mobile robot. But all that paled into insignificance when I encountered the university's 16-kilobyte digital computer mainframe and found books on artificial intelligence in the library. I adopted a schedule that allowed me to hack away at conventional AI techniques in the dead of night, when no one else was using the computer, and eventually there I was myself, in mainstream AI research at Stanford and MIT.

My second epiphany came much later. After bouncing back and forth between America's east and west coasts a few times, I became an assistant professor at MIT. The epiphany occurred during the January break in 1985, when I was on an extended family visit to Australia and Thailand. I had decided to build an intelligent mobile robot, and during this visit I had a lot of time to think about how I would program the robot to move around in the normal office environment.

My most recent work had been getting a robot arm to carry out simple assembly tasks. Much of the effort had gone into making sure that the robot had a complete understanding, ahead of time, of where all the parts were, how they were shaped, and the approximate tolerances of their positions and sizes. All this was too complex for the robot to sense with the current state of computer vision (and it still is), so every part of the assembly had to be modeled with precision. Like my earlier tic-tac-toe machine, the robot could operate only if everything went according to plan; it had no sense of an out-of-the-ordinary situation.

As I pondered how to make my new mobile robot operate in the relatively unstructured environment of an office suite, I realized that trying to keep track ahead of time of all possible circumstances the robot might encounter was impossible. And in watching ants at work and thinking about their neuroanatomy I realized that they must be operating in a very different way from how all our AI systems had been designed over the previous thirty years. I saw that there was a different way of organizing intelligence—one we had not yet begun to understand. That different way had the organism, whether biological or silicon-based, react to its environment anew at every step in time. It would give you a tic-tac-toe player that looked at a board with two X's on a diagonal and a blank third square and simply filled it in. It would give you a robot able to switch from following a corridor to turning in at the open door of a recharging room even if the corridor ended unexpectedly soon. And it would give you a robot brain much more adaptable than any of our current software. An insect loses a leg and immediately adapts; a person has a small stroke and is able to recover some lost abilities by using different cells within the brain.

Almost twenty years later I am still of the opinion that in order to get the sorts of performance we see in animal systems, we need to change the way we organize our intelligent systems. Robots based on my revelations that day in Thailand are now available in thousands of retail stores across the United States, and a robot programmed using the principles that were developed over the following few years has roamed at least one small corner of the surface of Mars. But the robots we know how to build today are still not nearly as robust as their animal models. There are flatworms, small ruffled inhabitants of coral reefs, whose brains can be transplanted from one individual to another; the brains regrow their connections and the creatures regain much of their abilities. More remarkably, a flatworm brain can be flipped over and placed in the recipient upside down or backward, and still much of the functionality is regained. This degree of flexibility and adaptability is currently not found either in the hardware of our computer chips or in our software systems.

So now I await my third epiphany. How can we organize the internal components of behavior-producing machinery within a silicon organism so that it can have the self-directed development and robustness we see in the simplest of animals? What metaphors are we misusing—the metaphor of information processing in neural systems, say—so that we are kept from a fuller understanding of how the behavior of biological creatures is ultimately generated by billions of localized molecular interactions?

As I look back at my life, I note a remarkable downward spiral of ambition. When I was eight years old, I wanted to build machines that were good at intellectual games that people play. In my thirties, I switched to trying to get the performance of an insect. Now I am after the secrets of a worm.

The Everyday Practice of Physics in Silver City, New Mexico

J. DOYNE FARMER

J. DOYNE FARMER, one of the pioneers of chaos and complexity theory, cofounded Prediction Company, in Santa Fe, a firm that does automated trading in financial markets based on mathematical algorithms. At Los Alamos National Laboratory, he did work in theoretical biology and founded the Complex Systems Group. He is currently McKinsey Professor at the Santa Fe Institute, where he applies ideas from physics and complex-systems theory to economics.

While I was growing up in Silver City, New Mexico, I never remember seeing a repairman enter our house. I don't think it would ever have occurred to my father to pay a specialist to do something he could do perfectly well himself. He had grown up on a peach farm in a remote part of the Ozark Mountains in Arkansas. After working as a machinist, as a dairy farmer, and in various other professions, he toured Europe via Normandy Beach, the Battle of the Bulge, and the march into Berlin. He then took advantage of the unprecedented opportunity of the GI Bill and became an engineer. As a result of all this, he knew how to do just about anything, and when anything around our house needed attention—the car, the hi-fi, the plumbing, the wiring, the kitchen cabinets—he fixed it.

From his example, I naturally assumed that building things was what people were meant to do. So I built lots of things, from soapbox-derby racers to a small house in the backyard. By the time I was ten or eleven, a friend and I had built at least ten treehouses or forts, to provide protection from the invaders who we knew were about to sweep into the neighborhood at any moment. When I was twelve, I saw James Bond, in *Thunderball,* escape into the air propelled by nothing but a rocket pack on his back, and my ambitions became more focused. That rocket pack was the coolest thing I had ever seen, and I decided I had to build one for myself. In the nearby public library, I read everything I could find on jet engines and rockets. I concluded that building a jet engine was probably beyond my abilities, but a smallish rocket pack was another matter. I schemed about how to build the chamber, what chemicals to buy, how to control it, how to avoid burning my legs off.

Around this time, I also joined the local Boy Scout troop. One evening a man named Tom Ingerson, in his twenties, came to a troop meeting and was introduced to us as a physicist. I was not exactly sure then what a physicist was, but I knew it was what Albert Einstein had been. Cool! Tom would be helping to run the scout troop. As it happened, he lived in our neighborhood, and afterward I walked home with him and asked him for advice on my rocket pack. He suggested some ideas of a more modest kind—like building a few small-scale models and testing them without myself attached. He had built rockets all through high school and had worked at the White Sands Missile Range. Then he asked me to take care of his cat while he went up to Boulder, to the University of Colorado, to defend his dissertation, in which he would present a new cosmological solution to the field equations of general relativity. To say that I was incredibly excited by this encounter would be an understatement.

A good deal of our lives is determined by chance, and in my life this is one of the best examples. If you know Silver City, New Mexico, pop. 7,000, you will immediately wonder why someone like Tom would be there. The main industry is the mining of copper, and my father was by then a superintendent in the mine. Tom was the entire physics department of Western New Mexico University, formerly New Mexico State Teachers' College. Western, as it was known, was not an institution celebrated for its high standards of scholarship.

The story of how Tom came to be in Silver City is a singularly American tale, and it had a huge influence on my becoming a scientist. Around the turn of the century, Tom's grandfather rode his high-wheeled bicycle all the way across the country to California. On the way back, his bicycle broke down in Texas. While he was fixing it, he fell in love with a local girl and landed a job as postmaster, a job he liked because it left him plenty of time to be an inventor. Tom's father grew up to be an engineer and worked at Bell Laboratories designing guidance systems for antiaircraft rockets.

On the other side of the family, Tom's Great-Uncle Jim was a surveyor and prospector. He spent a lot of time wandering around West Texas with his mule looking for gold. One day, in the Jefferson Davis Mountains, near the town of Alpine, he discovered what looked to him like the site of an old Spanish mine. A few years later, he was in Mexico on business when the Mexican Revolution shut down the government. In the ensuing confusion, he bribed an official to get admission to the government archives and there he found the records of a mine that appeared to be the one he had come across in the Jefferson Davis Mountains. According to the records, the Spaniards had used the local Indians as slaves until the Indians revolted and killed most of their oppressors. In desperation the Spaniards threw the gold down the shaft, blew it shut, and fled. There was no record of the gold having been recovered. Though Great-Uncle Jim made several stabs at locating the gold during the Depression, no amount of digging turned anything up.

Now to return to Silver City. As Tom was finishing graduate school at the University of Colorado, he decided to make use of his considerable skills in electronics and applied for jobs with several different companies. This was the mid-1960s, a time when

PhDs in physics were highly sought after; Tom assumed he would have no problem getting a job. To his surprise, he didn't get a single offer. Years later he found out why when the girlfriend of a friend who worked for Texas Instruments pulled his personnel file: He had been blacklisted. While a graduate student, he had worked to improve the physics teaching labs with Frank Oppenheimer, J. Robert's brother and an admitted past member of the American Communist Party. Frank Oppenheimer was Tom's best friend on the faculty, and Tom had naturally asked him for a recommendation. (Later, Frank Oppenheimer would found the Exploratorium, in San Francisco, which is probably the best science museum in the world.) The lack of a job created a pressing problem, since at age twenty-six Tom was still eligible for the draft. He had to find a job, fast. So he drew concentric circles on the map around Great-Uncle Jim's gold mine and applied to all the colleges that were within a day's drive. Fortunately, Western was desperate for a physics department. Tom sent them an application letter on a Friday in late August, and they called him Monday morning to offer him the job.

While Tom was defending his thesis up in Boulder, his cat had kittens, which was good because it gave me an excuse to come over nearly every day and hang out. Tom's house was amazing. It was totally squalid, which to me at the time suggested that any sort of life was possible. Everyone else I knew in Silver City was worried about what other people thought; it was wonderful to know someone who clearly didn't give a damn.

Tom returned from Boulder with one of his friends, known as Bog. Bog was weird, but he was also a lot of fun. He would sit on Tom's couch for hours, shooting at flies with rubber bands, and hitting quite a few of them. He only shot at flies on the wing, as stationary flies were not a sufficient challenge. Bog also liked to play chess, and as we were pretty well matched, we spent a lot of time doing that when he wasn't shooting flies. If he ever noticed that I wasn't an adult, he never mentioned it.

But by far the best part about going over there was listening to Tom. He was full of ideas, and he talked about everything under the sun. His intellectual passions spanned a huge gamut, from science to history to archaeology. In Tom's view, what was needed in life was a Great Cause, something grand that would be really worth doing. One of his favorite contenders was putting together a privately funded mission to Mars. NASA was wasting its time on relatively uninteresting enterprises like going to the moon, which everyone knew had no life on it. If NASA wasn't so hopelessly inept and inefficient, he said, they would already have had a mission to Mars in full swing. There was a lot of talk about their errors and the bad engineering directions engendered by Wernher von Braun. Tom played the piano quite well and liked to sing Tom Lehrer's "Wernher von Braun" (" 'Once the rockets are up, who cares where they come down / That's not my department,' says Wernher von Braun") or occasionally "The Elements" ("There's antimony, arsenic, aluminum, selenium, / And hydrogen and oxygen and nitrogen and rhenium"). He could do calculations in his head at blinding speed, and his brain was packed full of facts. The calculations involved exciting issues, such as estimates of how much gold might be in Great-Uncle Jim's elusive gold mine, what a minimal mission to Mars might cost if one did things intelligently, and whether the money we might get from the gold would be enough to fund such a mission.

Tom's books also opened up a new world for me. Other than a few books by Jules Verne, the Silver City Public Library didn't have much in the way of science fiction. But Tom had lots of it—classics by people like Isaac Asimov, Ray Bradbury, Arthur Clarke, and Fred Hoyle. My reading shifted to spaceships, robots, and psychohistory. There were frequent discussions with Tom about which stories had a plausible basis in science and which were just fantasy; these discussions taught me a lot of physics and opened my mind to fundamental questions about nature and metaphysics. A story about space travel would lead to a protracted discussion about relativity and the twin paradox, or possible power sources for spaceships and the pros and cons of nuclear power. Fred Hoyle's *The Black Cloud,* about an alien invader from interstellar space, spawned ongoing discussions about chemistry and what it means to be alive. Isaac Asimov's Foundation trilogy got us talking about what it means to predict something, and what is predictable, and what isn't.

Tom couldn't cope with the paramilitary nature of my local scout troop, so he decided to start an Explorer Scout post instead. Explorer Scouts are high school age, and though I still wasn't old enough I was allowed to participate in a lot of the activities of Explorer Post 114. We went backpacking frequently in the Gila Wilderness and the Superstition Mountains of Arizona and made road trips to Alaska and Yucatán. My friends and I began to make Tom's house our hangout. Meanwhile there were some big changes at home. My father was offered a better job working in a mine in Peru, and my parents decided to move. They were also having marital problems. Somehow I persuaded them to let me stay behind and move in with Tom, since I was already practically living with him anyway.

Then came motorcycles. I got a paper route, and to take care of my customers I had to have wheels. In New Mexico, a thirteen-year-old could get a license to drive a motorcycle as long as it was under five horsepower. So Tom took me to El Paso and we bought a 1963 Honda 50 Sport Cub. After only three weeks it threw a rod, and we took the engine apart in Tom's kitchen. I still vividly remember trying to separate the crankcase. We removed its bolts and pulled on the two halves, but nothing happened. So we pulled harder and harder and heard a loud *snap!!* when the remaining screw we had missed finally broke. What seemed like hundreds of little gears and springs and washers flew all over the kitchen. We carefully collected these and spent the next month staring at the wear patterns on the gears and thrust bearings to try to determine how everything went together. We assembled the engine again and again, until finally there were no parts left over. It worked fine and ran for many years, but Tom's kitchen never quite recovered. Before long, he bought a motorcycle himself, and then several of my other friends got motorcycles, and pretty soon Tom's yard was full of motorcycles and there was usually one disassembled in the kitchen.

We also took up electronics. I started by working on the wiring of my motorcycle and went on to more complicated things. Tom was a ham radio operator and an electronics expert; he made all kinds of equipment and had wire and solder and the other raw materials we needed. He taught my friends and me the basics and helped us design our own circuits. We built electronic ignitions for our motorcycles, homemade burglar alarms, and a crude prototype of a fax machine before anyone had heard of such a thing. I had a hard time waking up to do my paper route, so we built a contraption with a school bell loud enough to wake the dead, not to mention the neighbors.

And of course we made trips to the gold mine. These involved a long drive into West Texas, followed by several hours of hiking to get to the mine. It was a tantalizing place. There was a slide of loose rocks on a steep hillside, arranged roughly in a line heading straight down the side of the mountain. It did not look like a natural rock formation. At the bottom were the remains of a narrow dirt road leading down into an arroyo—a road clearly built with considerable hand labor and old enough to have mature trees growing in the middle of it. The theory was that the Spaniards extracted the gold ore from the mine, put it into bags, and slid the bags down the slide of rocks. They then carted the ore to the arroyo, where they crushed it into dust and panned it. We spent our time looking for Spanish artifacts (we never found any), collecting rock samples (all of which, the assayer consistently said, were just below the level of minable ore), and wandering around the top of the ridge trying to guess where the original mine shaft might have been amid the old holes dug by Great-Uncle Jim and his cronies during the Depression. Tom spun out ideas about how we might locate a deep but possibly dense concentration of gold. A normal metal detector wouldn't have nearly enough range, so we thought about radiation, powerful magnetic fields, and ways to dig deep holes cheaply in a very remote area. And of course we spent a lot of time thinking about how to mount our own expedition to Mars, once we finally found the gold.

Tom's example was not always easy to follow. He set a standard so high that it could be intimidating. I have since come to know some of the smartest people in the world, and now I see that Tom's intellect was (and still is) on a level with the best. I have not yet met anyone as good at solving complicated physics problems in his head. Tom was a master of the analytic method, which he applied to a wide range of interests. By the time I went to college, I was convinced that if I was to prove myself, then physics was the only true test—the only true path to deeper understanding of what makes the world tick. Furthermore, I knew that real physicists weren't just supposed to know about physics; they were supposed to be broadly educated and know a little—or, better yet, a lot—about everything.

Tom also taught me that science isn't just something scientists do in universities. It's a universal belief system, and it's also a way of solving everyday problems. Tom approached everything, from physics to politics, as a scientist, and showed me that there was no reason to let people intimidate you just because they were so-called experts. He encouraged me to take nothing for granted, to question everything. His style of thinking was subversively utopian; it made everything seem possible. His dreams infected me and all of my friends, and have motivated me through many difficult challenges in my life.

I am sometimes struck by the odd mixture of inexorable fate and purely random circumstance that has made me who I am. Events of historical importance can have far-reaching chance effects that intertwine with the natural unfolding of a single life. If it weren't for Adolf Hitler, we might never have had a GI Bill, my father probably wouldn't have become an engineer, and we would never have moved to Silver City. If it weren't for Senator Joseph McCarthy, Tom would never have taken the job at Western. If it weren't for both those consummate villains, we never would have met—and I might well have decided to become something other than a scientist. And who can say what that might have been?

The Math of the Real World

STEVEN STROGATZ

STEVEN STROGATZ is a professor in the department of theoretical and applied mechanics and the Center for Applied Mathematics at Cornell University. He is the author of *Sync* and the best-selling textbook *Nonlinear Dynamics and Chaos.*

What I like about the science of "complex systems"—that is, the attempt to understand the many occurrences in nature of spontaneous order, or self-organization—is that some of the major unsolved problems of science have this character. Architecturally, they involve millions of units—neurons, heart cells, players in an economy—all influencing one another through complex networks and via complicated interactions, and out of this you sometimes see amazingly organized states. You see synchrony.

The story of how I got interested in synchrony and cycles and the mathematics underlying them goes back to my freshman year at Loomis Chaffee School, in Windsor, Connecticut. On the first day of Science 1, our teacher, Mr. DiCurcio, told us to get down on our hands and knees and determine how long the corridor outside our classroom was. After five or ten minutes of painful crawling and laying down of rulers, I remember thinking to myself, "If this is what science is, it's pretty pointless—and dusty."

Fortunately I took to DiCurcio's second experiment a little better. "I want you to figure out a rule about this pendulum," he said, handing each of us a little toy pendulum with a retractable bob. You could make it a little longer or a little shorter in clicks, in discrete steps. We were each also handed a stopwatch and told to time ten swings of the pendulum, then click, lengthening the pendulum and noting how long it took for ten more swings, then click again, repeating the process. The point was to see how the length of the pendulum determined its period, the time for one swing to and fro. The experiment was supposed to teach us how to record a relationship between one variable and another on graph paper, but as I was dutifully plotting the period of the pendulum versus its length, it occurred to me after about the fourth or fifth dot that a pattern was starting to emerge. These dots were falling on a particular curve that I recognized because I'd seen it in my algebra class—it was a parabola, the same shape that water makes coming out of a fountain.

I remember experiencing an enveloping sensation of fear, then of awe. It was as if . . . this pendulum knew algebra! What was the connection between the parabolas in algebra class and the motion of this pendulum? There it was, on the graph paper. It was a moment that struck me, and it was my first sense that the phrase "law of nature" really meant something. I suddenly knew what people were talking about when they said that there was order in the universe, and that, more to the point, *you couldn't see it unless you knew math*. It was an epiphany I've never really recovered from.

I went from Loomis Chaffee to Princeton, where the path was a little bumpier. As a freshman, I began with linear algebra, the whiz-kid math course for students who had done well in math in high school. Our professor was to be John Mather. Mather today is one of Princeton's stalwarts, a renowned mathematician, but on that first day we couldn't tell if he was a professor or a grad student. He was intensely shy, with a long red beard, and far from striding into the classroom he seemed to slither along the wall. Except for the beard, he was practically invisible. Then he began,

"The definition of a field, *F*, is . . ." That was all. He didn't state his name, no "Welcome to Princeton," nothing. He just began with the definition at the beginning of linear algebra. And so it continued. Mather's course was a dreadful experience for me. For the first time in my life, I understood why people are terrified of math. He came close to discouraging me from ever wanting to be a mathematician.

The only reason I went on to become one was that my math course in sophomore year was with an exceptional teacher, Elias Stein, who is also still at Princeton and in 2001 won the National Medal of Science. It was a course in complex variables, which was a lot like calculus. I had always liked calculus in high school, and I suddenly felt as if I could do math again, whereas Mather's course in linear algebra had been a filter with a very fine mesh. Only certain students could get through the holes. What was supposedly being tested was your tendency to think abstractly: Could you come up with the sorts of rigorous proofs that a pure mathematician needs? That ability is the bread and butter of pure math. The truth is, I didn't have it in me; that wasn't my natural strength. What I really like is math applied to nature—the math of the real world. At the time, I didn't even know there was a thing called applied mathematics. Now it's what I do.

But there was still the home front to deal with. My parents had always encouraged me to be a doctor, and I had always resisted, because I knew I wanted to teach math. By my junior year at Princeton, my parents were urging me to take a few premed courses—some biology, some chemistry—and finally, even though it was getting to be pretty late to become a doctor, I agreed. My brother the lawyer convinced me with a persuasive argument that it was irrational to keep resisting: I wouldn't be committing myself to actually *being* a doctor, he said, and it wouldn't hurt to learn biology and chemistry. I accepted that—and it made for a hellish year. Along with my major in mathematics, I was taking freshman biology, with its lab; freshman chemistry, with another lab; and organic chemistry, for which freshman chemistry was a requirement. This is a lot for someone who's not much good in the laboratory. Nevertheless I liked what I was learning, especially the idea that DNA was a double helix and that this shape immediately indicated its function and explained how replication worked. So I thought I was perfectly content, and I even took a course to prepare for the MCAT (Medical College Admission Test).

When I came home for spring vacation, my mother took one look at my face and said, "There's something wrong. Something's really bothering you. What's wrong? How do you like school?"

"I like it," I said. "It's fine, I'm learning good stuff."

"You don't look right," she said. "You don't look happy. Something's wrong. What's wrong with you?"

I didn't really know. "Maybe I'm tired," I said. "I'm working a lot."

"No, something else is really wrong. What are you going to take next year? You'll be a senior."

Well, that *was* bothering me. "Being a premed so late, I'm going to have to take vertebrate physiology, some biochemistry, and all these premed courses. Plus I have a senior thesis in the math department. Which means that my schedule is going to be so full that I'm not going to be able to take quantum mechanics."

"Why does that matter to you?" she asked.

"I've been reading about Einstein since I was twelve years old," I blurted out. "I love Heisenberg, Niels Bohr, Schrödinger—I could finally understand what they're talking about! There won't be any more verbal analogies and metaphors. I'll understand what Schrödinger *did*! I've worked my whole life to get to this point, and I'm ready to know what the Heisenberg uncertainty principle really says, and instead I'm going to be in medical school, cutting up cadavers, and I'm never going to learn it!"

So she said, "What if you could just say, right now, 'I want to do math. I want to do physics. I want to take quantum mechanics. I'm not going to be a doctor. I want to be the best math teacher and researcher I can be?'" And I started to cry. It was as though a tremendous weight had been lifted. Then we were both laughing and crying. That was a moment of truth, and I never looked back. I'm very thankful that I had such good parents, and that I was able to find a passion by denying it for a while. Some people go through their whole lives without figuring out what it is they really want to do.

When it was time to produce a senior thesis—at Princeton, it's required—I wanted to pick something about geometry in nature, although I didn't know exactly what. My adviser was Fred Almgren, famous for studying the geometry of soap bubbles; he suggested a problem about the geometry of DNA. Could we understand, for instance, what it might be about DNA that allows it to unwind itself without getting tangled? Given that it's such a long molecule, you'd think it would get tangled occasionally—and that this would be deadly if it happened in the cell. What keeps it from doing that? While wrestling with that geometry problem (I never actually solved it), I also collaborated with a biochemist to propose a new structure for chromatin, which is a mixture of DNA and proteins that makes up our chromosomes; it's the next level of structure after the double helix. We know that the double helix gets wound around little spools of protein called nucleosomes, but no one knows how the nucleosomes themselves become arranged like beads on a string or how that structure is wound up to make chromosomes. My biochemical adviser and I ended up publishing a paper in the *Proceedings of the National Academy of Sciences.* That was exhilarating. I was now doing research in mathematical biology, math about real stuff, math about chromosomes.

That was when I realized that I wanted to be an applied mathematician doing mathematical biology. I went to Cambridge University on a Marshall scholarship, but I was put off by the traditional program—G. H. Hardy describes it in *A Mathematician's Apology*—known as the mathematical tripos. Cambridge has had the same kind of course since Newton; I was completely bored by it. One day I wandered into a bookstore across the street from my college, where I picked up a book with the unlikely title *The Geometry of Biological Time.* I had subtitled my senior thesis "An Essay in Geometric Biology," and I thought I had made up that peculiar phrase—geometric biology, the juxtaposition of geometry and biology, shape and life—and here was somebody using practically the same title. So who was this guy, Arthur T. Winfree?

I opened the book, and at first I thought he was nuts. The titles of his chapters were puns; he had data from his own mother's menstrual cycle. It was all about the cycles of living things. At the time, Winfree (who died on November 5, 2002) was not really on the map. He was a professor of biology at Purdue. I glanced at the book, put it back on the shelf, and found myself coming back a few days later to read some more. Eventually I bought it. Partly because I was bored and lonely, I started reading and underlining it every day. I fell in love with the vision of living things with many cyclic processes—cell division, heartbeat, rhythms in the brain, jet lag, sleep rhythms, all described by a single mathematics. This is what Winfree was proposing in his book, and it's what got me started in the direction of studying synchrony.

Here's a classic example of synchrony, from the natural world. Back in the time of Sir Francis Drake in the sixteenth century, there were persistent reports from the first Western travelers to Southeast Asia of spectacular scenes along riverbanks, where thousands upon thousands of fireflies in the trees would all light up simultaneously. These reports kept coming back to the West and were published in scientific journals, and people who hadn't seen the phenomenon couldn't believe it. Scientists said that this was just a case of human misperception, that we were seeing patterns that didn't exist, that it was an optical illusion. How could fireflies, which are not very intelligent creatures, manage to coordinate their flashings in such a vast and spectacular way?

One theory was that there might be a leader. But what would make one firefly special? It sounded ridiculous. We don't believe that there is a leader anymore, or that there might be atmospheric conditions that cause synchrony—for example, a bolt of lightning that startles every one of them and makes them start flashing at the same time. Synchrony occurs on nights that are perfectly clear. It was only in the 1960s that a biologist named John Buck, of the National Institutes of Health, and his colleagues figured out what was really going on, which is that the fireflies are self-organizing. They manage to get in step every night of the year, flashing in perfect time for hours on end, with no leader and no prompting from the environment, through what is essentially a mysterious process of emergence. The thinking now is that individual fireflies respond to the flashes of others—they adjust their own timers. Buck and his wife, Elisabeth, went to Thailand and collected bags full of fireflies, brought them back to their hotel room in Bangkok, and released them in the darkened room. The fireflies flitted around, crawled on the ceiling and walls, and gradually little pockets of two and then three and four fireflies began flashing in sync. Later lab experiments showed that by flashing a light at a firefly you could speed up or slow down its rhythm and make it flash a little sooner or later than it would otherwise.

You might ask why this matters. Who cares about fireflies? There are lots of reasons you should care: The first may be that all kinds of applications in technology and medicine depend on this same kind of spontaneous synchronization. In your heart you have 10,000 pacemaker cells that trigger the rest of your heart to beat properly, and those 10,000 cells are like the thousands of fireflies. Each one has its own rhythm—in this case, an electrical rhythmic discharge. Instead of communicating with light, they are sending electrical currents back and forth to one another, but at an abstract level they're fireflies—individual oscillators that want a periodic repetition of their state and can influence one another.

There are a host of medical and technological applications. The laser, one of the most practical gadgets of our time, depends on light waves in sync, atoms pulsing in unison, all emitting light of the same color and moving in phase, with the troughs and crests of the light waves perfectly lined up. The light in a laser is no different from light coming from a lightbulb, in that the atoms are no different; it's the coordination of the atoms that's different. The choreography is the difference, not the dancers.

What's so breathtaking about the phenomenon of synchrony is that it occurs at every scale of nature, from the subatomic to the cosmic. It's one of the most pervasive phenomena in nature, but at the same time one of the most mysterious from a theoretical perspective. We are used to thinking of entropy—the tendency of complex systems to get more and more disordered—as the dominant force in nature. People often ask me, "Doesn't synchrony violate that? Isn't it against the laws of nature for a system to become spontaneously ordered?" In fact there is no contradiction. The law of entropy applies to so-called isolated or closed systems, where there's no influx of energy from the environment. But that's not what we're talking about when we discuss living things. With systems that are far from thermodynamic equilibrium, all bets are off, and we see astonishing feats of self-organization, synchrony being the simplest such example. The same laws that give rise to entropy will also account for synchrony. It's just that we don't have a clear enough understanding of the thermodynamics of systems very far from equilibrium to see the connection—but we're getting there.

Recently I find myself wanting to learn more about cancer and what it is about the network of chemical reactions that goes awry in a cancerous cell. There are certainly some cases where a single gene may be responsible, but I don't believe that all cancers will be explained that way. Understanding oncogenes is a start, but not the whole answer. Again, it's about choreographies of proteins and genes—the steps not just of single dancers but of many, in movement together. Cancer is a dynamical disease that we won't understand through purely biological, reductionist thinking. It's going to take a combination of reductionism (to give us the data), new complex systems theory, supercomputers, and math. I would like to be part of that.

At Large in the Mountains

TIM WHITE

TIM WHITE is a professor in the Laboratory for Human Evolutionary Studies of the University of California at Berkeley, whose research emphasizes fieldwork designed to acquire new data on early hominid skeletal biology, environmental context, and behavior. He is the author of *Human Osteology*.

There were few clues, when I was a kid, that I would end up searching an African desert for the fossilized remains of long-forgotten ancestors. My parents were living in California, in a small cabin in the San Bernardino Mountains, near Lake Arrowhead, in August 1950, when I was born. The mountain cabin was my first home, but my earliest memories come from the next house we lived in, in Sky Forest, a small settlement along State Highway 18, a winding two-lane asphalt road that snakes along the crest of the main fault scarp of the San Bernardino Mountains, called by locals the Rim of the World. My father worked as a laborer for the county road department and then for the California State Highway Division. The new house was much larger and far more exciting for a kindergartner to explore. It was perched by itself on the edge of the San Bernardino National Forest, built precariously on a steep hillside just below the highway. Thousands of feet below the house, and an hour's drive down the hill, was the Inland Empire of Southern California. In those days, it was real estate notable for its groves of citrus trees, which stretched away in every direction. From the Rim of the World on a clear day, you could see Santa Catalina Island in the Pacific Ocean, just beyond Los Angeles. As Southern California filled with people and automobiles, there were fewer and fewer clear days. The orange groves disappeared. I grew up.

I have vivid memories of events in the house at Sky Forest. When I was six, the Shadow Mountain wildfire was ignited by an airplane crash. The conflagration threatened our little world. I remember the enormous size of that fire and the pungent smell of smoke from the burning chaparral as the flames raced up the slope toward our house. We evacuated for a week, but the house was spared. We moved back in and smelled the fire for months afterward.

Living on the edge of the forest opened the natural world to me in amazing ways. My younger brother Scott and I tried to domesticate mountain squirrels, a raccoon, several pigeons, blue jays, chipmunks, tortoises, turtles, and lots of snakes and lizards that shared the house and yard with the family dachshund. Rattlesnakes we were allowed to kill, but not to bring home alive. There were always terrariums and cages in the backyard, and most of our pets hibernated in the basement during the winter. Scott and I also had a donkey named Bimbo, who had been loaned to us by a neighbor. Wearing our Davy Crockett hats, we rode around and around the yard on that donkey with our toy rifles. We were heartbroken when Bimbo died from a rattlesnake bite. In the autumns, we would visit the nearby apple orchard and wonder at the fresh bear-claw marks on the trees, from which we collected apples that our neighbor's press would turn to cider. At night, my brother and I would listen to the hot Santa Ana winds, or the cold winter storm winds, blowing through the pine trees and out over the twinkling lights in the valley far below. Those winter storms dropped thick blankets of snow on the mountains and made for great sledding. I was such a small and light kid that I could sled atop the hard crust of snow that formed each night. One morning I was doing just that and got a little too close to the edge of the road. The slope below approached vertical. Were it not for a few thin but strong branches of a young oak tree that I managed to grab hold of, I would have had a fast ride to oblivion hundreds of feet below. I never told my parents about this, for fear of losing the sled.

The years at Sky Forest passed quickly. On Labor Day the tourists and vacationers went back to their homes in the valley, and soon afterward I went off to elementary school, several miles away, with the few other children who lived year-round in the mountains. On Memorial Day, the tourists and vacationers would reappear. Summers were difficult for the family whenever my father was sent to the desert to repair state roads damaged by flash floods. He would be gone for two or three weeks at a time, and his return was eagerly anticipated by my brother and me, because he would always bring us something, dead or alive, from the desert. It was a fascinating but distant world I discovered through the tortoises, lizards, strange plants, and unusual rocks he brought home. He worked his way up through the ranks of the state highway system and became a heavy-equipment operator. Working on the roads, he learned about places in the back country, the more arid northeastern slope of the mountains that was undeveloped national forest crossed only by trails and dirt roads. On the weekends the family would drive out to explore that country in all of its natural, historical, and geographic dimensions. We all learned together on those trips.

My father had an abiding interest in California's history, and although the history of the San Bernardino Mountains was not very deep, it was rich. The Indians had come first, then the Spanish missionaries and ranchers, the Mormon settlers, and the loggers and miners. Each set of inhabitants left its mark: The pioneers became settlers and the settlers became developers. As the sawmills fell silent, the developers eventually built dams to create lakes at Big Bear and Little Bear (Lake Arrowhead). By the time Scott and I were growing up, they had built a golf course and a country club.

Even as a child I was fascinated by this history. Each week I would beg our parents to explore for more evidence of it. We made trips to the Mojave Desert and the Pacific. After four years at Sky Forest, we moved closer to Lake Arrowhead, to a house that was built for us in Cedar Glen. Now our family life revolved around the lake. We had a small boat with an outboard motor, and we water-skied on summer evenings when the lake was as flat as glass. Other summer evenings were spent on less enjoyable but more profitable, character-building endeavors, making extra money by cutting lake-weed, which rich summer vacationers paid to have removed from their docks and beaches. My father would dive below the lake's surface with a large scythe, cutting the weed near its base. As it floated to the surface, my job was to rake and pitchfork huge masses of the wet, cold, crawfish-infested stuff into piles on the beach, which we would later haul away. Winter brought additional opportunities. While my father worked long shifts plowing the snow from the highways, my brother and I would shovel people's driveways for money or put tire chains on the cars of the tourists, who sometimes arrived at their resort cabins in Bermuda shorts and usually without gloves. There wasn't much to spend our earnings on, so we saved most of it. A vague notion of college lay on the distant horizon.

On the summer weekends, when the water-skiing wasn't so good because of all the tourists' speedboats, the family would take the outboard across to the north shore of the lake, where there were still stretches of unpopulated beach. While my brother swam and my parents sunbathed, I hunted lizards in the rocks, listened to LA Dodgers games on a transistor radio, and read many books. My favorites were the Time-Life books on natural history my grandmother had given us. The text and pictures made accessible the natural world that surrounded and fascinated me. I read two of them so often that the pages fell out. One was *Evolution* and the other was *Early Man,* the latter by Clark Howell, whom I would join at Berkeley years later. *Evolution* explained the principles that governed the natural world I saw around me. Everything made sense in this Darwinian explanation, including the desert tortoise that had urinated in my brother's lap as soon as we brought it into the car on one of our trips. And *Early Man* told me how stone tools were made, how prehistoric remains and artifacts were preserved and brought to light, how it was possible to discover the many lost worlds of the past. I remember being unbelievably excited when my father brought home a prehistoric ground-stone bowl that had been exposed during road widening at Baldwin Lake, near Big Bear. I begged him to take me to the place so that we could find more.

When I was nine years old, Mary Leakey found a cranium in Olduvai Gorge. *National Geographic* brought this and subsequent discoveries into our living room. They were a little distant for a mountain kid to relate to. I was more interested in baseball and live reptiles than in anything archaeological. Africa was another planet. Like many others at that age, I was fascinated by dinosaurs. Kid-friendly books about dinosaurs were not as plentiful as they are today, but that didn't matter—I wanted to find dinosaur fossils more than I wanted to read about them. Unfortunately, the San Bernardino Mountains are mostly igneous rock, and there were no fossils within bicycling distance. One weekend I persuaded my parents to drive us out to the Mojave Desert near Barstow to search "the fossil beds." None of us knew what we were doing. My brother was never very happy in the desert anyway, and coming home empty-handed just made matters worse.

We took family trips to the Grand Canyon, and camped in the backcountry of the San Bernardinos. We tracked mountain lions along Deep Creek and watched beavers make dams in the streams. But this natural world of the mountains was shrinking. The permanent resident population was growing. Development was virtually unchecked outside the boundaries of the national forest. Thousands of new houses linked by hundreds of new roads appeared. More and more people moved to the mountains. My father became a highway foreman and finally a superintendent whose territory extended from the Inland Empire to the Colorado River.

Sometime during junior high school, I visited the school counselor to discuss careers. I said I would like to find dinosaurs. Not even my mother, let alone the counselor, took this career choice seriously, and the counselor proposed marine biology as a more practical option. Nobody in my family had ever gone to college. In school, we had been warned repeatedly by our teachers that almost no one from our mountain had successfully made it through the University of California. When I finally did go to college, at UC Riverside, I was ready to fail, too. It was a pleasant surprise to get C's in freshman chemistry and physics. Not until I was a junior did I decide I could do better academically.

In high school I began to read and understand more about archaeology. The backcountry was the perfect place to pursue that interest. No archaeological survey had ever been done there, but the San Bernardino County Museum was excavating at a place called Rock Camp, or Indian Rocks—a seasonal encampment where centuries of prehistoric use had produced many bedrock mortars and a deep midden. A friend of mine had excavated there, and together we explored the Deep Creek drainage in an effort to find other sites. We ended up locating dozens of them and made a substantial artifact collection that the county museum later accessioned, along with all of our meticulously kept records. I taught myself how to read a topographic map, how to recognize and make stone tools, how to get around in the wilderness. And I bagged many snakes along the way.

I enrolled at UC Riverside in 1968, figuring on a career in marine biology. That never materialized. It was Scott's heart that was set on the sea, and he eventually moved to Hawaii, where he lives today, centered in Earth's largest ocean. My heart remained with the desert. My father had bought me a used 1966 Chevrolet pickup truck, which I still own, as a graduation present. This vehicle opened wider horizons. I could now go deep into the Mojave Desert in search of new old things, and sidewinders. I was nearly thrown out of the university for keeping snakes in my dormitory room. At Riverside I majored in biology, focusing on terrestrial field biology and learning from Professor Wilbur Mayhew, an icon in the field biology of Southern California.

Archaeology continued to interest me throughout four years of college, but I didn't take an anthropology course until I was a junior. When I did, I walked out of my first class, a discussion section in introductory archaeology taught by a graduate student bent on reifying a site-classification system he'd learned from our textbook. Real archaeological sites, which I already knew something about, weren't patterned that way, and I told the instructor so. When he became authoritarian, I collected my books, excused myself for having wasted his time, and walked out of the classroom, earning raised eyebrows and a few smiles from my fellow students. I added a second major in anthropology after better experiences (mostly reading) in physical anthropology and fieldwork in archaeology and vertebrate paleontology. I made it all the way through UC Riverside and went on to get a PhD from the University of Michigan in 1977. Then came a visiting lectureship at Berkeley, which eventually turned into a professorship.

My parents tell me that as a kid I was always on the go, often alone, always looking for something interesting, committed to the outdoors and to learning everything about the world of nature from rocks to rattlesnakes, aardvarks to zebras. I grew up in what I thought was a huge mountain world, a world that grew progressively smaller and more comprehensible.

So what made the kid choose a life in science? It was the freedom. My parents never pushed me in any career direction. For this freedom I cannot thank them enough. I inherited their skepticism (they were dead set against religious orthodoxy), their fascination with things historical and natural, their curiosity. The opportunity for a kid to grow up in the mountains came only because my parents were young and willing to take a risk, to seek a better and more interesting life. I was the luckiest kid in the world, a privileged passenger along for the ride of his life.

Today when I go back to the mountains, I see that much of the world I learned from as a kid has already disappeared. Forested hillsides of my childhood are now covered with huge sporadically occupied houses behind gated driveways. The wildlife is mostly gone and so is the solitude. I suppose the spatial, cultural, and temporal dimensions of most kids' worlds increase in the same basic ways—but I was lucky to be in the mountains when life was easier than it had been for the pioneers, and better than it is now.

The Making of a Scientist

V. S. RAMACHANDRAN

V. S. RAMACHANDRAN is director of the Center for Brain and Cognition; a professor in the psychology department and the neurosciences program at the University of California, San Diego; and an adjunct professor of biology at the Salk Institute. He is the author of *Phantoms in the Brain.*

What is the single most important quality that suits you for a career in science? People often say "curiosity," but surely that can't be the whole story. After all, *everyone* is curious to some degree, but not everyone is destined to be a scientist. I would argue that you need to be obsessively, passionately, almost pathologically curious. Or, as Peter Medawar once said, you need to "experience physical discomfort when there is incomprehension." Curiosity needs to dominate your life.

Science is a love affair with nature—a love affair that has all the obsessive qualities, the turbulence, the passionate yearning that one commonly associates with romantic love. But where does this yearning come from? To some extent, it is probably an innate personality trait. But more important, it arises from your early associations. I realized a long time ago that the best formula for success is to be around people who are passionate and enthusiastic about what they do, for there is nothing more contagious than enthusiasm. I was very fortunate in this regard. In the British school in Bangkok that I attended through fourth grade, I had some exceptionally gifted science teachers—Mrs. Vanit and Mrs. Panachura—who would give me chemicals to take home and do experiments with. Again, at Stanley Medical College, in Madras, my biology professor, Dr. Rao, would pretend he was the monk Gregor Mendel breeding pea plants in the monastery garden, in order to convey to us the excitement of how these seemingly arcane experiments created the whole new science of genetics. And at home, there was the influence of my science-minded uncles and my brother Ravi, whose passion for poetry and literature, especially Shakespeare, rubbed off on me. Science has a great deal more in common with poetry than most of us realize; both enterprises involve unusual juxtapositions of ideas and a certain romantic vision of the world.

It helps, too, to have parents like mine, who constantly goad you to excel and who stimulate rather than stifle your natural curiosity. Knowing of my interest in science, my mother brought me seashells and other zoological specimens (including a tiny seahorse) from all over the world and helped me set up a chemistry lab under our staircase. When I was eleven years old, my father bought me a Carl Zeiss research microscope. More important, they planted two mutually incompatible ideas in my head (and I'd recommend that any parent reading this do the same): first, that I was the chosen one, the very best; and second, that I was never good enough for them. It's a surefire formula guaranteed to turn your child into a success, albeit perhaps a neurotic one.

I was interested in science from about the age of eleven. I remember being a somewhat lonely child, socially awkward. But I always felt companioned by nature; perhaps science was my retreat from the social world, with all its arbitrariness and mind-numbing conventions. Whether I was collecting seashells or geological specimens, I felt that the natural world was my private playground—my own parallel universe, inhabited by Charles Darwin, Georges Cuvier, Thomas Henry Huxley, Richard Owen, Jean-François Champollion, William Jones. For me those people were more real—certainly more alive—than most "real" people I knew. This escape into my own private world made me feel special and privileged, rather than isolated or "weird." It allowed me to rise above the tedium and monotony, the humdrum existence that most people call a normal life. (One recalls Bertrand Russell's remark: "Real life to most of us is a perpetual compromise between the ideal and the real, but the world of pure reason knows no compromise, [allowing] one, at least, of our nobler impulses to escape from the dreary exile of the actual world.")

Students, colleagues, and reporters often ask me why I became interested in the brain, and when. It's not easy to trace the lineage of one's intellectual preoccupations, so I'll start by making a list of things that have always fascinated me, both professionally and as an amateur. I love archaeology, particularly the study of ancient Indian history, archaeology, and art. Together with my postdoctoral colleague Eric Altschuller, I have even tried cracking the Indus Valley script. (Needless to say, we were not the first to attempt this. It is the world's last major undeciphered script and has been intensively studied by professional linguists as well as thousands of crackpots and wannabes like us.) For example, earlier studies had established that the symbol 0 means "goat" or "sheep" in Sumerian and is pronounced "udu" or "audu." I was struck by the fact the South Indian / Dravidian (Tamil) word for "goat" is also *audu,* and, amazingly, when I looked through a museum catalog of Indus seals, there were seals with exactly the same symbol, with a picture of a goat right next to them! I realized that this was unlikely to be a coincidence and suggested that all three great civilizations (Sumerian, Indus Valley, Dravidian) may have shared the same dialect and script five thousand years ago and that the symbol 0 is pronounced "*audu*" in the Indus language and means "goat." Suddenly a long-lost alphabet sprang to life! Eric and I had fun doing this, but I should add that we have yet to be vindicated by Indology pundits.

I am perpetually trying to cultivate orchids—unsuccessfully. I love anthropology and ethnology. I am obsessed with paleontology and collect fossils in the field. One of the most memorable was a 30-million-year-old skull of Oreodon, an extinct herbivorous mammal, which I excavated on a cliff face in South Dakota; I still savor the feeling of knowing that mine were the first human eyes to gaze into its bony orbits. I like dabbling in comparative anatomy, teratology (the study of biological malformations), and natural history; I find endlessly fascinating the idea that the tiny bones inside our ears, which we mammals use for amplifying sounds, originally evolved from the reptile jawbone. When I was a teenager in India, I wrote papers about gastropod taxonomy. I was passionate about inorganic chemistry and often mixed chemicals together just to see what would happen. A burning piece of magnesium ribbon could be plunged into water and would *continue to burn* by extracting the oxygen from H_2O. Another passion was botany: I tried placing various sugars and amino acids inside the "mouths" of Venus's-flytraps to determine what would trigger them to clamp down and secrete digestive enzymes. And when I failed to water a plant, I noticed that instead of the whole plant wilting progressively, a single leaf would fully dry up first. Was this the plant's way of conserving its water for as long as possible? I enjoyed collecting bugs and dreamed of becoming India's foremost entomologist; I experimented with ants to see if they would hoard and consume saccharin as they did sugar. Would the saccharin molecule fool their taste buds as it fools ours?

Now, at first these diverse topics seem to have nothing in common. But they are all characteristically Victorian preoccupations. This insight has often made me feel like an anachronism (or, as my grad students say behind my back, an old fossil) in the age of high-tech science.

Science flourishes best in an atmosphere of complete freedom and financial independence. No wonder it reached its zenith during times of great prosperity and patronage of learning: in ancient Greece, where logic and geometry first emerged; in the golden age of the Guptas in India, around the fifth century A.D., when the number system, trigonometry, and much of algebra as we now know it were born; and during the Victorian era—the era of gentlemen scientists like Darwin and Lord Henry Cavendish. Today we have the tenure system and federally funded grants whose patronage many of us are lucky to enjoy, but unfortunately they sometimes have the effect of punishing the visionary and rewarding the sycophant. (As Sherlock Holmes told Watson, "Mediocrity knows nothing higher than itself; it requires talent to recognize genius.") The solution most of us adopt is to apply for funds to support our more down-to-earth, "safe" projects and use some of the money to pursue more adventurous research on the side.

But I want to probe deeper and ask why Victorian-style science is so appealing to me and to many of my colleagues. After all, no one would deny that modern science is immensely successful. Perhaps one reason is simply my impatience with gadgetry. I get uneasy when the gap between the raw data and the conclusion is too big, because it generates an irresistible urge to massage the data. Partly because of this obsession with technology, a lot of research these days tends to be methodology- and gadget-driven rather than problem-driven. The result is an extremely boring enterprise—what you might call sciencemanship rather than science. I have sometimes wondered whether—contrary to Edison's remark that "genius is ninety-nine-percent perspiration" (often quoted to schoolchildren to make them tolerate drudgery)—the importance and impact of a scientific discovery might be inversely proportional to the time and effort that went into it. This isn't always true, of course, but it's true surprisingly often. The greatest biological discovery of the twentieth century, the structure of DNA, was made by Francis Crick and James Watson with bits of wire and plastic, in less than six months. Which is not to deny the long period of intellectual incubation that must have occurred before they actually tackled the problem. Let us bear in mind that an important or "fundamental" scientific problem is not necessarily more difficult to solve than a trivial one; after all, Nature isn't conspiring against us to hide her secrets. So, given our finite life span, it's best to focus on the former.

In the Victorian era, science was a great adventure. Whole new worlds were opened up. Just think of the excitement of the early explorers when they confirmed the first rumors of the existence of great apes—creatures on the brink of becoming human. Or when Alfred Russel Wallace first saw the birds of paradise. Or when a strange beast called the giraffe was discovered that had the same number of cervical vertebrae as a sparrow but freakishly elongated. Or when Darwin discovered a new species of orchid with a foot-long flower and predicted the existence of a species of moth with a correspondingly long proboscis for sucking its nectar—a prediction subsequently confirmed. Or when anthropologists first saw the Pygmies—people under four feet tall—living less than a hundred miles away from the Watusis, the tallest race in the world. Or when Michael Faraday moved a magnet into a coiled wire and found that it generated an electric current, thereby linking electricity and magnetism.

Of course, technology drives science as surely as ideas do; just think of the microscope or the telescope, not to mention the bubble chamber and the computer. But the eyes behind the telescope are equally important. Many people peered through telescopes before Galileo, but he was the first to look at the sky instead of terrestrial objects and that made all the difference. The last two decades have seen the emergence of marvelous new brain-imaging devices, fMR (functional Magnetic Resonance) and PET (Positran Emission Tomography), which promise to do for brain research what the telescope did for astronomy. What brain research needs, though, is a few Galileos and Faradays, not the voyeuristic neophrenology—the "Gee whiz, let's see what lights up" approach—that now dominates the field. I also think the time is not yet ripe for theoreticians like James Clerk Maxwell (though one of my theoretically minded colleagues at UC San Diego, Terrence Sejnowski, may prove me wrong).

Science is the most fun when it's in its infancy—when its practitioners are still driven by curiosity and it hasn't become just another nine-to-five job. Unfortunately this is no longer true for many of the most successful areas of science, such as particle physics or molecular biology. It is now commonplace to see a paper in *Science* or *Nature* with thirty authors. This "assembly-line" approach takes much of the joy out of the practice of science and is one of two reasons I instinctively gravitate toward old-fashioned *Geschwind*ian neurology, where it's still possible to ask naïve questions starting from first principles—the kinds of simple questions that a schoolchild might ask but are embarrassingly hard for experts to answer. Neurology is a field where it's still possible to do Faraday-style research—mere "tinkering"—and come up with surprising answers with far-reaching implications. Many of my colleagues and I see it as an opportunity to revive the golden age of such pioneers in the field as Jean-Martin Charcot, John Hughlings Jackson, Henry Head, Alexander Luria, and Kurt Goldstein.

The second reason I chose neurology is much more obvious and has to do, again, with curiosity. As human beings, we are more curious about ourselves than we are about anything else, and neurology is a discipline that takes you right into the heart of the problem of who we are. I got hooked on it about ten years ago, after examining my very first patient in medical school. He was a man with a pseudobulbar palsy—a kind of stroke—who alternately laughed and wept uncontrollably every few seconds. His behavior struck me as an instant replay of the human condition. Was this mirthless joy and crocodile tears, I wondered? Or was he actually *feeling* alternately happy and sad, as a manic-depressive might, but on a compressed timescale?

Many people who embark on a scientific career do so in the hope of becoming famous, and I am no more immune to such vanities than any of my colleagues. But at least such considerations don't occupy center stage in my mind anymore, because there are two things I know for sure: I have more fun doing science than I ever expected to (so much so that one colleague with a strong Protestant work ethic once asked me suspiciously, "Can it really be science if you are having so much fun?"), and many of the experiments I have done on perception and neurology have influenced the thinking of at least some of my colleagues in these fields. In the final analysis, only these two questions matter when you look back over your life: How much impact have I had? And how much fun?

What I Want to Be When I Grow Up

DANIEL C. DENNETT

DANIEL C. DENNETT is University Professor and Austin B. Fletcher Professor of Philosophy and director of the Center for Cognitive Studies at Tufts University. He is the author of, among other books, *Consciousness Explained, Darwin's Dangerous Idea, Kinds of Minds,* and, most recently, *Freedom Evolves.*

I had lots of adventures when I was a kid, but none of the kind that would prepare me for a life on the edge of science. None of my many mentors was a scientist, and I had no scientific epiphanies until I was in graduate school. I came from a family of historians and English teachers and doctors, and it was assumed that I would do something in the arts or humanities. My father was a Harvard historian (with a brief stint at Clark University), an expert on Islamic history, and a fluent Arabic speaker. He turned his talents to the nation's use during World War II, when he was recruited to be a secret agent in the OSS, headquartered in Beirut, where we lived in diplomatic splendor thanks to his cover job as cultural attaché at the American Legation.

Few four-year-olds ever have a pet gazelle, but for a while I did. I named it Babar, after my favorite character in literature at the time. It leapt around in our rather large and high-fenced garden, and was the gift of some Bedouin potentate, perhaps the same man I visited with my father one day in the desert and returned home with my ears pierced. I was a blond child, an unprecedented marvel in the experience of the Bedouins, warranting this special treatment. My mother was horrified and pulled out the strings at once, but the little knife-blade scars remain. Our nursemaid, a young Armenian girl named Mary, loved to encourage my crayon drawings, and our "chauffeur," a nice young Lebanese man, had few driving tasks to occupy his time, so he often helped me with my first construction projects: a little wooden table and chair, and a kite, which he showed me and my older sister how to fly on the hill in Shemlan, where we went in the summer when Beirut was too hot to live in comfortably. When my father wasn't off on some "educational project" (read: mission), he presided over many festive gatherings of his Beirut friends and contacts, colleagues at AUB (the American University of Beirut), British and French diplomats, expatriates. Not having known any other way of life, it was only in later years that I came to appreciate how exotic and exciting this all was. I went to the nursery school run by the education department at AUB, where I spoke Arabic and French with my classmates and amused my parents' friends who asked if I was going to school yet. "Yes," I would reply, "I go to AUB."

My father died in an airplane crash in the mountains of Ethiopia in 1947, when I was five. My younger sister, Charlotte Dennett, who was born in Beirut within weeks of Father's death, is currently writing a book on his OSS exploits and the situation in the Middle East during the scramble for oil in the 1940s. From her I've recently learned details about his death that she's dug up using the Freedom of Information Act and the reminiscences of aging spooks. I'll let her tell that tale in a year or so.

After Father's death, Mother moved the family back to Massachusetts, where she took a job as high school social studies editor at Ginn and Company, the Boston textbook publisher. She had been an English teacher in her home state of Minnesota before venturing to Beirut to teach at the American Community School, and she knew how to make a sentence work. Hardly a day went by when she didn't bring home some tale of a woebegone author—a history professor at a major university, more often than not—whose lamentably limp and fuzzy exposition she had had to revise, or simply replace with her own tough, clean prose. Her authors were always grateful, sometimes embarrassingly appreciative of what they owed her, but while some of them got rich on royalties (even in those days, a history or social studies textbook was a gold mine if the large states adopted it), the most she got from them was a few bottles of bourbon or a few toys for us kids.

While Mother took over the breadwinner role, commuting to Boston every day, our housekeeper, a somewhat older woman whom we children named Cookie (to her initial dismay, for she was much, much more than a cook), filled the mothering role, providing the daily discipline and oversight and loving us as her own. She and Mother often disagreed, and my sisters and I participated vigorously in their suppertime conversations. We had a rule at the dinner table: The only reason for being excused in the middle of the meal was to go look up some point of contention in the *World Book* or other reference work. There were frequent consultations. Our home was full of books and magazines, and after school, when I wasn't reading I was drawing, constructing, rebuilding things in my basement workshop (for the large-scale carpentry and metalwork) or in my attic bedroom, where I kept all the small tools for sculpting and drawing, along with my gigantic chest of several Erector sets and the extraneous machine parts that might come in handy someday. I shunned all the diagrams and instructions that came with the Erector sets, preferring to strike out on my own and make something original. From the age of five, I was fascinated with taking things apart and repairing things, but the question of whether I might want to become an engineer never came up. In our circle, it was a prospect about as remote as becoming a lion tamer.

On the eve of going off to Beirut to be a spy, Father had extracted from his boyhood friend, Sherman Russell, a promise to take care of "Ruth and the kids" if anything happened to him. Sherm, a lifelong bachelor, was true to his word: He had helped us move back to the States and found a house for us to live in. He then became my surrogate father for many years, and I used to embarrass Mother by asking her when she was going to marry him. I adored this eccentric man, with his many unusual talents and tastes. He was an expert horseman with a passion for foxhunting, and he was for a time Master of Hounds for Lady Molly Cusack in Ireland. He would embark from Boston's Logan Airport for Shannon wearing his "pinks" and carrying his saddle. An outrageous snob, and not afraid of being conspicuous, he lived rather frugally in his boyhood home on his family money and never had a regular paying job. He helped people every day—a pillar of the community who could always be counted on in an emergency. I had no interest in polo or foxhunting, though, and while Sherm was always on hand Christmas morning to help me set up the new electric trains or build something truly spectacular out of my newest Erector set, he didn't, in fact, teach me much of anything. (One Christmas morning, while he and I were engrossed in building a giant drawbridge on the living-room carpet, his aged mother, who had accompanied him, turned to my mother and said, "Isn't it nice that my little boy and your little boy can get together on Christmas!")

I went to summer camp in New Hampshire every year, and when I was about thirteen I discovered jazz, under the spell of Ed and Dick Lincoln, camp counselors who played drums and vibraphone. I switched piano teachers, and learned where I could buy (under the table) the chief tool of the trade: the fabled fakebook, a three-ring notebook containing over a thousand "lead sheets" of jazz and popular songs. The fakebook violated copyright law egregiously, but it was so useful in standardizing the keys, chords, and lyrics of the standards that every jazz pianist had one. Mine got so much use that I had to go through it every few years and reapply those little glue-on reinforcement rings around all the holes, a lengthy procedure in a book with almost four hundred pages. My dream was to be a lounge lizard, to play smoky ballads of exquisite sweetness and sophistication with somebody like Michelle

Pfeiffer leaning misty-eyed over my shoulder. Actually, since this was the middle of the fifties, my fantasy ideal was probably closer to Doris Day.

The high point of my life as a jazz musician would occur one night in Paris in 1961, when I went with my Harvard classmate Ronald Brown, a wonderful jazz pianist, to Le Chat Qui Pêche. It became a jam session about three in the morning, and we both sat in with Chet Baker. I played just one number and sensibly retired; Ron played till dawn. (A year or so later he killed himself, for reasons I've never learned, and the memory of his playing that night remains frozen as my ideal of the musical life I might almost have had.) Through high school and the first year of college, I played in combos of one sort or another and did quite a lot of arranging for small bands and singing groups, but I eventually realized that I would have to work too hard to pump up my musical talents to a professional level.

Perhaps, then, I would instead become a great artist. I was an Eagle Scout, and my scoutmaster, Paul Butterworth, was a commercial artist with a deft line that I envied and tried unsuccessfully to emulate in the cartoons I drew for the troop's weekly newspaper, which we mimeographed. (Try drawing a line with any zing using a stylus on a waxy stencil! Paul could do it.) I was more your Garry Trudeau kind of cartoonist—a pathetic line, but good ideas—and when Paul said to me one day, "Danny, you have great ideas, and that's the main thing in art," he said something that has reverberated in my head ever since.

Maybe, then, I would be a sculptor, like my heroes Henry Moore or Constantin Brancusi—sculptors get to nibble away at their lines till they look just right. I had learned the art of whittling at summer camp, thanks to another camp counselor, Garland Thayer, Appalachian pied piper and craftsman extraordinaire, and soon added all manner of chisels and other sculptor's tools to my kit. I branched out into soft stone, sheet metal, and other materials. I sculpted throughout high school and college, and in 1961, after my sophomore year, I spent the summer as an apprentice of sorts in the studio of Pietro Consagra in Rome (it was on my way to Rome that Ron Brown and I had our session in Paris). Consagra had recently won the Venice Biennale and was the intellectual leader of a vigorous group of sculptors in Rome; the Basaldella brothers, Afro and Mirko, and Arnaldo Pomodoro are the ones who stick in my mind. I was entranced by the lifestyle in Rome, but also somewhat oppressed and outraged. This was the Rome of Cinecittà and I was wandering around the edges of the crowd portrayed in *La Dolce Vita*. One night at supper I was introduced to Federico Fellini, of whom I had never heard. "What do you do?" I asked. "Cinema director," he replied. Oh. When I left Rome at the end of the summer to return to college, I vowed I would never go back to that decadent, painful city, but a few summers later, when my young wife and I were driving south through Italy in search of a place to hole up for the summer while I wrote the first draft of my Oxford dissertation, Rome beckoned and held us. We rented a furnished flat in Trastevere, and soon all was forgiven. But unlike the previous summer in Athens, when I'd obtained a large hunk of Pendelic marble and whacked away on a piece every morning, I was now committed to being a philosopher, with sculpture only a sideline.

In my first year at Winchester High School I had had two wonderful semesters of ancient history, taught by lively, inspiring interns from the Harvard School of Education, and I poured my heart into a term paper on Plato, with a drawing of Rodin's *Thinker* on the cover. Deep stuff, I thought, but the fact was that I hardly understood a word of what I read for it. More important, really, was that I *knew* then—thank you, Catherine Laguardia and Michael Greenebaum, wherever you are—that I was going to be a teacher. The only question was what subject. I spent my last two years of high school at Phillips Exeter Academy, and there I was immersed in a wonderfully intense intellectual stew. It was the kind of place where the editor of the literary magazine had more cachet than the captain of the football team, where boys read books that weren't on the assigned reading list, where I learned to write (and write, and write, and write). I was admitted to the legendary George Bennett's creative writing class in my senior year, and I churned out hundreds of pages on my trusty Olivetti Lettera portable typewriter (just like the one my ancient history teacher, Michael Greenebaum, had had—*cool!*). But none of those pages was philosophy yet. Perhaps I would be a novelist.

I knew that I would be a teacher, but not—I was sure—a science teacher. In the first lab of my high school chemistry class, we learned how to use our Bunsen burners to bend glass tubing, and I managed to hand a very hot glass bend I'd just made to my teacher, who made the mistake of grabbing it, right on the bend. My biology teacher in ninth grade was a junior high football coach whose idea of pedagogy was to give us multicolored anatomical diagrams of clams, frogs, and worms, on which we were to write all the proper names of their various parts. I was also taught a rudimentary sketch of the hierarchy of kingdom, order, class, phylum, genus, and species, without a whisper about why they were organized in this uptight, hierarchical way. It was the Dewey Decimal System applied to living things, apparently. What next, memorizing the phone book? If this was biology, they could have it.

In spite of this, I had a subscription to *Scientific American* for several years from the age of about twelve, and every month I'd pore through it, usually just looking at the diagrams and pictures and reading the captions. I loved the ideas, but never thought of becoming a scientist. It was only in graduate school at Oxford that I began to take seriously the idea that I might participate in scientific exploration. As a result, I'm an autodidact—or, more properly, the beneficiary of hundreds of hours of informal tutorials on all the fields that interest me, from some of the world's leading scientists. Lucky me. But I wasn't a kid when I fell in love with science. I just felt like one. I wonder what I will be when I grow up.

write and write and write and write [illegible] I was dimly aware of the fog [illegible] and [illegible] hundred [illegible] pages on my [illegible] [illegible] had [illegible] [illegible] was [illegible]

I [illegible] that I would be a teacher [illegible] was not [illegible] ence teacher, in the [illegible] high school [illegible] [illegible]

In [illegible] [illegible]

The Gift of Solitude

JUDITH RICH HARRIS

JUDITH RICH HARRIS is a writer and developmental psychologist. She is the author of *The Nurture Assumption: Why Children Turn Out the Way They Do*, which was a finalist for the Pulitzer Prize in 1999.

On February 12, 1957, a nine-year-old girl in Virginia wrote a letter to President Eisenhower, expressing her support for school desegregation. Forty-five years went by before the letter writer—now a distinguished historian—saw her letter again. As she explained recently in her alumnae magazine, it came as a surprise and not an altogether pleasant one. She remembers herself as not particularly religious as a child; she has had no religious affiliation as an adult. But her letter to the president gave a religious reason for desegregating the schools. "Jesus Christ was born," the future historian said, not just to save the white people but "black yellow red and brown" as well. The adult historian had no recollection of writing those words, though she remembered writing the letter. But she remembered having written a *different* letter—one that was more in line with her current reasons for opposing segregation.

No, the nine-year-old girl was not me; I'm a scientist and a writer, not a historian, and I was nineteen, not nine, in 1957. And no letters from me are likely to make an embarrassing reappearance in presidential archives. In fact, my early life is almost entirely undocumented: My family moved around a lot, and nonessential paperwork was jettisoned. My memories of childhood will therefore go uncontradicted—and unsupported.

The malleability of memory (see also Steven Pinker's essay in this volume) is the first reason why autobiographies should be taken with a grain of salt. As the historian, whose name is Drew Gilpin Faust, observed, "We create ourselves out of the stories we tell about our lives, stories that impose purpose and meaning on experiences that often seem random and discontinuous." Or, as the psychologist Elizabeth Loftus put it, "Memory is a creative event, born anew every day. You fill in the holes every time you reconstruct an event in your own mind." You fill them in on the basis of your current perspective, which may differ from the perspective you had when the event occurred.

The second reason we shouldn't take autobiographies too seriously is that people view both past and current events through the lens of their culture. The stories that impose meaning on experiences are a product of the cultural mythology of their time and place. Developmental psychologist Jerome Kagan provided a good illustration. In his 1998 book *Three Seductive Ideas,* he contrasted the autobiographical writings of Alice James (sister to Henry and William) with those of the writer John Cheever. Both were subject to spells of depression, but Alice James, writing in the latter half of the nineteenth century, "believed with the vast majority of her contemporaries that she had inherited her nervous, dour mood," Kagan reported, whereas Cheever, writing in the latter half of the twentieth, "assumed that his bouts of depression were due to childhood experiences . . . the conflicts that he imagined his family had created."

The lesson conveyed by James vs. Cheever is that people don't necessarily know why they turned out the way they did and most simply accept the explanation approved by their culture. Ironically, this lesson was lost on the very person who provided it. When I questioned the culturally approved explanation in my book *The Nurture Assumption* and found it unsupported by the evidence, it was Jerome Kagan who wrote (in a 1998 essay in the *Boston Globe*) that my conclusions must be wrong because they didn't agree with "biographical memoirs" he had read.

The final reason for taking biographical memoirs with a grain of salt has to do with how the mind works. As Steven Pinker explained in his book of that name, "The mind is organized into modules or mental organs, each with a specialized design that makes it an expert in one arena of interaction with the world." Though some of these mental organs—for example, the one that specializes in recording people's names, faces, and distinctive characteristics—operate on a level readily accessible to the conscious mind, most do not. A mental mechanism that operates below the level of consciousness doesn't leave conspicuous traces of its activities in memory. The reason we don't know why we turned out the way we did is that some of the mental mechanisms responsible for how we turned out didn't keep us informed about what they were doing. We cannot recall what was never encoded in the first place.

My memories are no more trustworthy than anyone else's. Bear that in mind while you read the remainder of this essay (and the other essays in this book).

Autobiographers generally describe their parents as either good or bad, helpful or hurtful. Not many parents seem to fall in the middle. Mine were in the middle. They were not intellectuals; in most respects they were ordinary people. My mother did well in high school, but her family couldn't afford to send her to college. My father's family was better off financially, but he was a poor student, probably dyslexic. He dropped out of high school after his father died.

His father died of an autoimmune disorder called pernicious anemia. My father, too, had an autoimmune disorder: ankylosing spondylitis, a form of arthritis that affects the spine. For most of his adult life he held no job; he was a crotchety invalid, often in pain. The four of us—my family also included a younger brother—lived on his disability insurance and on his inheritance from his parents, wisely invested. My father's health was the reason we moved around so much, back and forth between Arizona and the East Coast. By the age of thirteen, I had lived in seven different houses or apartments and attended eight different schools. I was kicked out of the first one, a nursery school; they said I wouldn't follow orders. "Tell me about it" was my mother's reaction.

I was active and headstrong and lacked the things that little girls were supposed to have in those days: grace, charm, dimples. To make matters worse, I was always among the youngest and smallest in my class in school and one of the few who wore glasses. I had learned to read early, and when my parents entered me in kindergarten the school switched me to first grade, despite the fact that I was small for my age and socially immature. School nonetheless went reasonably well, until my family made one move too many and I landed in a fourth grade classroom in Yonkers, New York. I became a social reject; none of my classmates would talk to me. During the four years we lived in Yonkers, I was an outcast.

For solace, I turned to reading. My parents didn't approve; "You're ruining your eyes!" they told me. One of my favorite books was *Little Women*. I identified, of course, with Jo, the budding writer, but it never occurred to me to aspire to be a writer myself. A naturalist, perhaps? I had an interest in wildlife, but I didn't know for a long time that one could make a career out of it. What I expected to be when I grew up (if I grew up) was a wife and mother. In the meantime, I mothered my dolls and my pets. My parents were permissive when it came to pets; I kept animals of all kinds. In addition to the usual dogs and cats, there was a lizard, a horned toad, turtles, a rabbit, a kangaroo rat, hamsters, hooded rats, a parakeet, and a baby robin that we raised successfully. As an undergraduate at the University of Arizona, working as a lab assistant in zoology, I brought home a boa constrictor one day.

But I'm getting ahead of myself. Something happened in the school in Yonkers, in fourth or fifth grade, that left an impression. A couple of teachers, acting as roving reporters, came to our classroom one day to collect material for the school newsletter. The question they asked our class was, "What is the height of your ambition?"— a pretentious way of framing the question. They were expecting answers like "To be a movie star" or "To pitch for the Yankees," but I shouted out "Five-foot-two." Everyone laughed. But when my smart-ass reply appeared in the newsletter, it had suffered an editing: The quote attributed to me was "To be five-foot-two." The teachers hadn't gotten my joke, I realized. Teachers aren't that smart! Teachers don't know everything!

I never quite achieved the height of my ambition, by the way. Five-foot-one-and-a-half was the highest I got.

My sojourn as an outcast ended when I was twelve and my family moved back to Arizona. There was a dramatic improvement in my academic performance as well. On my first day in the new school in Tucson, the teacher gave us a test in biology. As usual, my parents had made the move in the middle of the school year, so I hadn't done whatever work the test was based on. But the teacher told me to take it anyway. Afterward the other kids asked me how I did. When I showed them the grade on my paper, they were impressed. "Wow," they said, "you must be a brain!" And so I became an academic achiever and began hanging around with the brainy kids in school. In high school I belonged to the literary set; the editors of the school newspaper and magazine were my friends. I contributed humorous pieces to the magazine and wrote headlines for the paper.

But being a member of the literati of Tucson High wasn't what turned me into a writer: It was my health. As I mentioned, autoimmune disorders run in my family. From the time I was born, my immune system was out to get me. I've managed to remain alive for sixty-five years because its attention span is short and also thanks to modern medical and surgical interventions. The early symptoms were varied and mysterious. In infancy I had rashes, in childhood unexplained fevers and joint pains. When the usual viruses came around, I always got sicker than anyone else. By adolescence I required daily supplements of thyroid hormone. Autoimmune thyroiditis was unknown then, but fortunately the hypothyroidism it produces was detected and treated.

The healthiest period of my life was late adolescence and early adulthood. I graduated from college and entered graduate school, the department of psychology at Harvard. When Harvard kicked me out—they told me they doubted I would ever make a significant contribution to psychology—I married one of my fellow grad students. We reared two daughters. When my younger child started school, I got a part-time job as a research assistant at Bell Labs, in Murray Hill, New Jersey. I was just starting to think about going back to grad school when my immune system began to inflict serious damage. It struck first at the joints of my back, later at a variety of organs. The back problem was painful; for the first time, I appreciated why my father had been so crotchety. I couldn't sit comfortably or stand for long, so I spent a good deal of time in bed. Watching television, reading novels, doing double-crostics, constructing double-crostics—all got boring after a while.

Then a friend—Marilyn Shaw, an assistant professor of psychology at Rutgers—gave me something better to do. A paper she had submitted to a journal had been turned down. She thought it might stand a better chance if the writing were improved and offered me the job of editing it. I had "a way with words," she said. Her opinion was based on a classified ad I had once helped her write when she was trying to find a home for a dog. Random events can change the course of one's life. In my case, it was a classified ad about a dog—the shortest writing job I've ever had.

The package Marilyn sent me included not only the existing draft of the paper but also the data it was based on, from an experiment, using human subjects, on a kind of information processing known as visual search. I soon realized that the paper needed more than a cosmetic makeover: It needed an explanation of the data. It needed, to be precise, a new mathematical model. Neither of the existing models of visual search, which Marilyn's experiment was designed to test, provided a good fit for her data.

With the rich lode of Marilyn's data to play with, I became a scientist. I spent a year happily punching numbers into a little hand calculator, working out the details of my model. It included a power function, one of the few things I learned at Harvard that proved useful later on. B. F. Skinner's "laws of behavior," I had by then discovered, were not of much use in child rearing, though they worked well enough on our dogs and cats.

As the model improved, so did the paper. I produced draft after draft. The final version was accepted by a journal; I was senior author by then. From my bed I wrote a grant application and Marilyn received a grant to test the further predictions made by the model. Its predictions were gloriously upheld, but Marilyn didn't live to see our second paper published. She died of cervical cancer in 1983.

By then I had become a writer. That was Marilyn's doing, too. She was asked to contribute a chapter to an introductory psychology textbook and hired me to do the writing. Then the publishers of the textbook asked me to write a chapter on the brain and nervous system. Then they asked me to coauthor a textbook on developmental psychology. It was unknown territory for me, but I said yes and learned as I went along. I wrote the first edition of *The Child* in bed, in longhand, a spiral-bound notebook propped on my knee; my older daughter did the typing. During the years I spent writing and updating *The Child* (it went through three editions), I wasn't a scientist, just a writer. There is little scope for originality in writing a textbook for undergraduates. All I was doing was parroting what the recognized authorities were saying. At the time, I believed them.

I stopped believing them abruptly, and became a scientist again, on January 20, 1994. It was after a year spent reading widely in diverse areas of psychology, in preparation for writing a new, biologically oriented textbook on human development. Though my reading had included some of the seminal papers in behavioral genetics and evolutionary psychology, I had something more conventional in my hands when the lightning struck: an article on adolescent delinquency. My epiphany (described in chapter 12 of *The Nurture Assumption*) changed my view of childhood. It made me question almost everything I had read in developmental psychology and almost everything I had written. It gave me the idea for a new theory of personality development. I abandoned the textbook. Nine months later I submitted a paper to the *Psychological Review.* My paper laid out the evidence against the existing theories of personality development and presented the new theory. The article was published in 1995; it won an award. I began work on *The Nurture Assumption.*

Currently I am working on a new book about the evolution of personality. Though my health has not improved, I can sit comfortably for hours now; the spiral-bound notebook has long since been replaced by a computer. My grab bag of symptoms has been diagnosed as "an overlap syndrome of lupus and systemic sclerosis," two autoimmune disorders that can each affect a variety of organs. Over the years there have been problems with my joints, skin, blood, endocrine system, digestive system, and central nervous system; now my immune system has focused its sights on my heart and lungs.

What made me become a scientist and writer? Genetic factors are no doubt involved. I seem to have been born with the predispositions to love reading and to thumb my nose at authority. But what environmental factors were influential?

Not my parents. They were not "role models" for me, and I'm not the kind of person they were hoping I'd become. They would have liked me to be more like other girls.

Not my teachers, either. I can't name a single teacher, from nursery school through graduate school, who had an important influence on me, aside from the two who didn't get my joke.

Which brings me to my peers. According to my theory, humans are innately motivated, as a result of their evolutionary history, to ally themselves with a group of others like themselves—for children, that means the peer group—and to tailor their behavior to that of their group. This process, called socialization, makes children more similar in behavior to their peers. But there is another process, operating at the same time, that makes children less like their peers: differentiation within the group. The members of a group differ in status, or are typecast by the others in different ways, which widens the personality differences among them.

Strangely enough, I seem to be deficient in the motivation to ally myself with a group. Is it a result of having been an outcast for four years as a child? Or was I born this way? After all, people do vary in the strength of innate motivations such as maternal instinct and sex drive. Perhaps the reason I was rejected by the kids in Yonkers is that I wasn't sufficiently motivated to conform to their standards. Come to think of it, that may also be the reason why I was kicked out of Harvard. Both rejections had beneficial consequences in the long run. If I hadn't been an outcast for four years, I wouldn't have become an introvert. I might have fulfilled my parents' dreams and turned out just like the other girls. And if I hadn't been kicked out of Harvard, I wouldn't have become a maverick, nailing my disputations to the door of the orthodoxy. But why did it take me so long to find my hammer?

What's unusual about my story is that my periods of greatest creativity occurred when I was home alone. Most science nowadays is collaborative. Most scientists and thinkers thrive in places where they can toss around ideas with others. But it wasn't until after my paper was accepted by the *Psychological Review* that I began exchanging e-mail with scientists and thinkers in the academic world. Until then, I was flying solo. Even now, I've met few of my e-mail colleagues in person.

The influence of my peers in college and graduate school seems to have been largely negative; it was many years before I began to question Harvard's assessment of me. Perhaps my doubt that I could make a significant contribution as a scientist was due to the way my peers had typecast me. Though in junior high I was pegged as a "brain," I lost that distinction when my peers became brainier themselves. The problem was that I didn't look like someone who would ever accomplish anything important. I was small and childish-looking. People wanted to pat me on the head. They didn't take me seriously.

But how you look doesn't matter when you work alone and communicate with others only in writing. It was solitude that eventually enabled me to become myself.

www.randomhouse.co.uk/vintage